Greening the Road:
Policies and Technologies for Lower-Emission Cars in the EU

Jacob

Contents

CHAPTER 1

Introduction

In the European Union (EU) the transport sector is responsible for nearly 30 % of the total CO_2 emissions. The CO_2 emission proportion of the road transportation on the transport sector is 72 %, whereby about 61 % can be related only to passenger cars [1]. Therefore, one goal of the EU is to reduce the CO_2 emissions caused by the transport section by 60 % until 2050 compared to the level of 1990. However, in contrast to the EU's CO_2 emission reduction goal, the CO_2 emissions for the passenger transport sector increased by 3.7 % between 1995 and 2018, although the CO_2 emissions per transport capacity[1] decreased by about 9 % [2]. At the same time, the passenger car traffic increased by about 14 % between 1995 and 2018 [3]. As consequence, the CO_2 emissions for passenger cars and light commercial vehicles are regulated by the EU. In 2015 the threshold for CO_2 emissions of a passenger car was defined by $130\,\mathrm{g}\,CO_2/\mathrm{km}$. And for 2021 the target value is only $95\,\mathrm{g}\,CO_2/\mathrm{km}$ [4].

In the last 50 years many arrangements have been made to reduce fuel consumption, pollutants and CO_2 emissions. In Germany in 1968 the German Road Traffic Licensing Regulations were supplemented with rules for the measurement procedures for vehicle air pollution. In this regulation a driving cycle for a dynamometer bench was specified, to verify the average emission of pollutants in a busy urban region after a vehicle cold start [5]. The regulation was then implemented in 1970. Subsequently, this regulation was updated permanently and in 1992 the New European Driving Cycle (NEDC) was defined. In 2013 the validity of the regulation of 1970 ended and was repealed by Regulation No. 715/2007, which is valid until now [6]. The used type approval procedure is illustrated in Figure 1.1, schematically.

The first step to get a valid type approval is the determination of the road load of a vehicle, for example with the coastdown method carried out at a proving ground. After

---

[1]The transport capacity means the mileage multiplied with the amount of transported passengers. The unit is passenger kilometres (pkm).

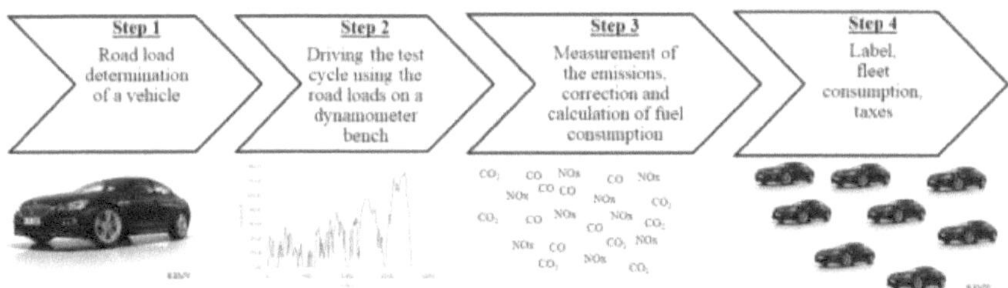

Figure 1.1: Type approval procedure (adapted by permission from Springer Nature Customer Service Centre GmbH: Springer Nature, 16. Stuttgarter Symposium Automobil- und Motorentechnik [7], 2016, and 18. Internationales Stuttgarter Symposium [8], 2018).

that, the vehicle drives a test cycle (e.g. the NEDC) on a dynamometer test bench, using the road load of the vehicle previously determined. Whilst the vehicle is driving the chosen test cycle, its emissions in the exhaust are measured and finally t he f uel c onsumption is determined. Concluding, this data is used to determine the CO_2 labels, the vehicle taxes and the fleet consumption [7]. To a pproach t he f uel c onsumption d etermined o n t he test bench to the customer experience, a new test procedure called Worldwide harmonized Light vehicles Test Procedures (WLTP) was created. For the evolution of the included driving cycle called Worldwide Light-Duty Driving Test Cycle (WLTC) worldwide driving data were accumulated and evaluated [9].

This clarifies t hat t he d etermination o f t he r oad load o f a vehicle (Step 1 i n F igure 1.1) is an important step for the whole type approval process. In addition, it has to be considered that the absolute value of the road load depends on the vehicle defined by its form, weight and equipment versions. Having a look at the today's portfolio of BMW (Bayerische Motoren Werke), which includes 59 different vehicles, it is obvious that the process of the road load determination requires great effort. In addition, the BMW Group not only consists of the brand BMW, but also of the brands MINI, ROLLS ROYCE and BMW Motorrad. Moreover, for each vehicle series a variety of derivates, model lines and equipment versions exists and each possible combination results in a vehicle specific r oad l oad, which has to be determined. However, the allowed coastdown method is strongly influenced by environmental conditions such as atmospheric temperature, wind velocities and wind directions. Therefore, test tracks are necessary, which have stable environmental conditions. These test tracks are usually located in Southern Europe, which results, amongst others, in additional costs for vehicle and passenger transport to the test tracks. But with the introduction of WLTP it is now possible to determine the road load of a vehicle not only using the coastdown method on a test track, but also with the so-called wind tunnel method. This method allows to separate the road load into two parts and determine these parts separately at two test benches - wind tunnel and flat belt dynanometer [10].

At the beginning of this **book** in September 2016, the flat belt dynamomter of the

BMW Group was just in the phase of comissioning. So there was no practical experience available about the test bench itself and also not about the wind tunnel method. Furthermore, at this moment only one publication of Rohde-Brandburger et al. was known which investigates the CO_2 potential of vehicles on a flat belt dynamometer of Volkswagen [11]. Therefore, the first part of this study investigates the sensitivities of the determination of the road load using the wind tunnel method in combination with the flat belt dynamometer. But there are still two test benches (wind tunnel and flat belt dynamometer) necessary. Considering this, the second part of this study investigates, whether it is possible to determine the total road load of a vehicle also in a wind tunnel. The benefits are obvious: The road load of a vehicle can be determined in one single test bench with only one measurement. This further reduces the costs for logistic processes and personal demand. Additionally, the road load determination is independent of environmental influencing factors such as temperature, wind velocity and direction and is therefore similar to the wind tunnel method always verifiable for authorities.

CHAPTER 2

Theoretical foundations

This chapter consists of the following parts:

- Theoretical description of the road load

- Composition of the road load of a vehicle

- Methods of road load determination

- Explanation of the cycle energy demand and the difference in cycle energy demand and a description of the investigated driving cycles

- Exemplary calculation of the road load using a fictional vehicle

- Explanation of the error calculation according to GUM (Guide to the expression of Uncertainty in Measurement)

2.1 Theoretical description of the road load

In this section, the road load F_i of a vehicle is theoretically derived. It is assumed that a vehicle is accelerated to a minimum velocity of $135\,\mathrm{km/h}$ and then is rolling, while the vehicle transmission is placed in neutral. During the deceleration the velocity v_i over the time t_i is measured, which is illustrated in Figure 2.1 (a).
Based on this information the road load F_i can be calculated making the following assumptions:

- A force, and thus also the road load, can be calculated by multiplying the mass (m_{eq}) with the acceleration (a_i).

- The road load (F_i) curve can be described as a second-order polynomial (see Figure 2.1 (b) and Equation 2.2).

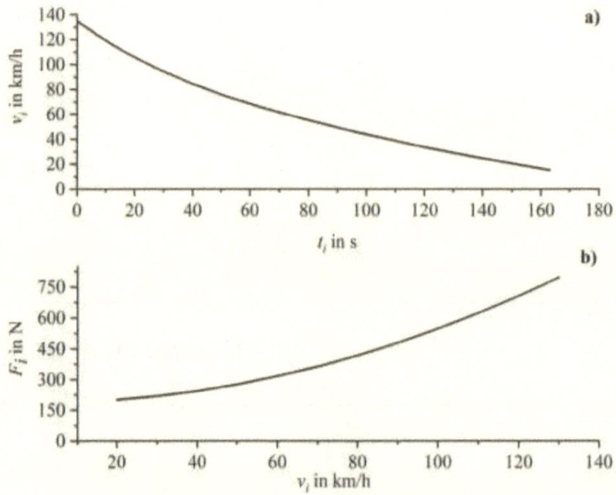

Figure 2.1: (a) Vehicle velocity v_i over the elapsed time t_i during the deceleration of a vehicle; (b) Resulting road load F_i over the vehicle velocity v_i.

Considering the first assumption

$$F_i = -m_{\text{eq}} \cdot a_i = -m_{\text{eq}} \left.\frac{dv}{dt}\right)_i \tag{2.1}$$

in combination with the second-order polynomial curve characteristic

$$F_i = f_0 + f_1 \cdot v_i + f_2 \cdot v_i^2 \tag{2.2}$$

the road load can be expressed as [12]:

$$m_{\text{eq}} \left.\frac{dv}{dt}\right)_i = -\left(f_0 + f_1 \cdot v_i + f_2 \cdot v_i^2\right) \tag{2.3}$$

The equivalent vehicle mass m_{eq} does not only consider the vehicle mass m_{Veh} but also the inertial effect of the rotatory components (see subsection 2.2.5 or Equation 2.36). Afterwards Equation 2.3 is integrated [12]:

$$\int_{t_0}^{t_i} dt = -m_{\text{eq}} \int_{v_0}^{v_i} \frac{dv}{f_0 + f_1 \cdot v + f_2 \cdot v^2} \tag{2.4}$$

And with the integration approach given in [13]

$$\int \frac{1}{ax^2 + bx + c}\,dx = \frac{2}{\sqrt{4ac - b^2}} \cdot \arctan\left(\frac{2ax + b}{\sqrt{4ac - b^2}}\right) \tag{2.5}$$

which is valid for $4ac - b^2 > 0$ [13], Equation 2.4 yields in [12]:

$$t_i = \frac{2 \cdot m_{\mathrm{eq}}}{\sqrt{4 f_0 f_2 - f_1^2}} \left[\arctan\left(\frac{2 f_2 v_0 + f_1}{\sqrt{4 f_0 f_2 - f_1^2}} \right) - \arctan\left(\frac{2 f_2 v_i + f_1}{\sqrt{4 f_0 f_2 - f_1^2}} \right) \right] =$$

$$= \frac{1}{\Theta_1} \left[\Theta_0 - \Theta\left(v_i \right) \right] \quad (2.6)$$

with

$$\Theta_0 = \Theta\left(v_0 \right) = \arctan\left(\frac{2 f_2 v_0 + f_1}{\sqrt{4 f_0 f_2 - f_1^2}} \right), \quad (2.7)$$

$$\Theta_1 = \frac{1}{2 m_{\mathrm{eq}}} \sqrt{4 f_0 f_2 - f_1^2} \quad (2.8)$$

and

$$\Theta\left(v_i \right) = \arctan\left(\frac{2 f_2 v_i + f_1}{\sqrt{4 f_0 f_2 - f_1^2}} \right) \quad (2.9)$$

Finally, Equation 2.6 is transposed to [12]:

$$v_i = \frac{\sqrt{4 f_0 f_2 - f_1^2}\, \tan\left(\Theta_0 - \Theta_1 t_i \right) - f_1}{2 f_2} = \frac{2 m_{\mathrm{eq}} \Theta_1 \tan\left(\Theta_0 - \Theta_1 t_i \right) - f_1}{2 f_2} \quad (2.10)$$

where:

F_i	is the road load of a vehicle at time step i in N;
f_0	is the constant term of the road load coefficient in N;
f_1	is the coefficient of the first order term of the road load coefficient in $\mathrm{N}/(\mathrm{m/s})$;
f_2	is the coefficient of the second order term of the road load coefficient in $\mathrm{N}/(\mathrm{m/s})^2$;
t_0	is the start time of the deceleration in s;
t_i	is the time step i in s;
v_0	is the velocity at the start time t_0 in $\mathrm{m/s}$;
v_i	is the velocity at time step i in $\mathrm{m/s}$;
m_{eq}	is the equivalent vehicle mass in kg;
a_i	is the acceleration at time step i in $\mathrm{m/s}^2$;
Θ_0, Θ_1	are constant terms;
$\Theta\left(v_i \right)$	is a variable depending on v_i;
a, b, c	are constants;
v	is the integration variable of the velocity in $\mathrm{m/s}$;
t	is the integration variable of the time in s.

Afterwards, the road load coefficients of the vehicle, which describe the road load curve (shown in Figure 2.1 b), can be calculated with the least square method using Equation 2.10 [12]. However, as Equation 2.10 is a non-linear equation, the road load coefficients f_0, f_1 and f_2 can be only estimated using an iterative process [12, 14]. For example, to solve this kind of optimization problem, the Generalized Reduced Gradient (GRG) method can be used, which is also implemented in Microsoft Excel. For more information about this algorithm see for example [15].

2.2 Composition of the road load of a vehicle

The road load of a vehicle is a force in the opposite direction of the vehicle motion. It has to be overcome by the power and energy of the powertrain to accelerate the vehicle and to keep the vehicle at constant velocity. The road load of a vehicle is defined as the sum of the following components [16, 17, 19]:

- Aerodynamic drag F_{Air}

- Rolling resistance F_{Roll}

- Drivetrain losses F_{Drive}[2]

- Climbing resistance F_{Climb}

- Inertial resistance F_{Inertial}

A detailed description of these individual resistances is given in the following subsections.

2.2.1 Aerodynamic drag

If a body like a vehicle is moving, it has to displace the surrounding fluid (in this case air). Due to the friction of the fluid, a resistance to movement develops, known as aerodynamic drag. This aerodynamic drag F_{Air} is an irreversible energy loss due to the flow losses. The aerodynamic drag is usually expressed according to Equation 2.11 [16, 18]:

$$F_{\text{Air}} = c_{\text{w}} \cdot A_x \cdot \frac{\rho_{\text{Air}}}{2} \cdot v_{\infty}^2 \tag{2.11}$$

where:

F_{Air}	is the aerodynamic drag in N;
c_{w}	is the aerodynamic drag coefficient;
A_x	is the frontal area of the vehicle in m^2;
ρ_{Air}	is the air density in kg/m^3;
v_{∞}	is the inflow velocity in m/s.

---

[2]The drivetrain losses are here defined as the losses which occure between the disengaged clutch and the wheels.

In the following, a further differentiation of the aerodynamic drag is given [16, 18]:

- **Form drag**: This is the resistance of the smooth basic body without attachments and surface design and without lift forces. It contains both pressure and friction components.

- **Cooling air drag**: This resistance is a result of the flow through the engine compartment and the flow of cooling air to the brake systems, gear transmission and catalysts.

- **Induced drag**: If a fluid is flowing around a body, lift forces and forces downwards occur and a system of wake vortices is generated, which increases the aerodynamic drag. Thus, this resistance results due to a different pressure distribution at the body under and upper side.

- **Roughness drag**: The roughness drag can be understood as the resistance of attachments and the surface design, e.g. underbody with wheel suspensions, wheels and exterior mirrors.

- **Interference drag**: However, to calculate the total aerodynamic drag, it is not possible to sum up the above described, individual resistances. Instead, the interaction between these factors has also to be considered. For example, attachments not only have an influence on the roughness drag but also on the vehicle form drag. The interference drag accounts for these influences.

In Figure 2.2 the percentage allocation of the aerodynamic drag into its components form drag (dark grey), cooling air drag (grey), interference drag (white) and roughness drag (light grey) is illustrated. Furthermore, the roughness drag is divided into its influencing parts wheel, underbody and exterior mirrors. It is shown that the largest proportions are related to the form drag and the roughness drag. It has to be considered that the measure that influences the form and the cooling air drag also has an impact on lift forces and forces downwards and therefore on the induced drag. Therefore, the induced drag is not listed separately in Figure 2.2. A more detailed explanation of the above listed resistances is given in [18] and [20].

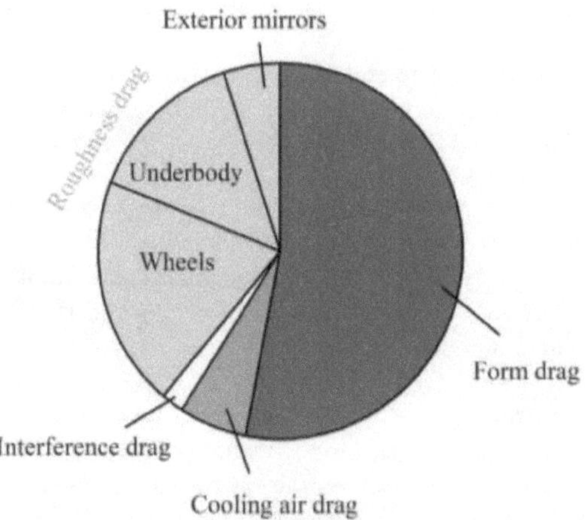

Figure 2.2: Percentage allocation of the aerodynamic drag by the single components form drag (dark grey), cooling air drag (grey), interference drag (white) and roughness drag (light grey), which is further divided into its influencing parts wheels, underbody and exterior mirrors (own representation based on [18]).

Further influencing factors on the aerodynamic drag

Until this point, the aerodynamic drag is determined with a constant air density. However, due to changes of ambient conditions such as air temperature and pressure also the density changes. Furthermore, additional headwind, tailwind and sidewind as well as c_w dependency on wind velocity not only have an influence on the absolute value of the aerodynamic drag coefficient c_w, but also on the lift and down forces of a vehicle which lead to ride height changes and therefore also to c_w changes. These complex interactions are explained in more detail below.

Ambient atmospheric temperature and pressure

Neglecting the influence of the air humidity, the ambient atmospheric density ρ_Air depends on air temperature and pressure and can be calculated using the ideal gas equation [21]:

$$\rho_\mathrm{Air} = \frac{p_\mathrm{Amb}}{R \cdot T_\mathrm{Amb}} \qquad (2.12)$$

where:

p_Amb	is the atmospheric pressure in Pa;
R	is the specific gas constant with the value of $287.0\,\mathrm{J/kg \cdot K}$;
T_Amb	is the ambient atmospheric temperature in K.

Headwind, tailwind and sidewind

Considering the headwind, tailwind and sidewind, the inflow velocity in Equation 2.11 can be expanded to read:

$$v_\infty = v \pm v_\mathrm{w} \qquad (2.13)$$

where:

v	is the vehicle velocity in $\mathrm{m/s}$;
v_w	is the wind velocity in $\mathrm{m/s}$.

If only headwind or tailwind is present, the aerodynamic drag increases or decreases accordingly. The fraction of the sidewind to the wind velocity along the longitudinal axis of the vehicle can be calculated with the aid of the inflow angles of the sidewind [22]. In the case of sidewinds an inflow velocity v_res results from the vehicle velocity v, the wind velocity v_w and the related sideslip angle θ as shown in Figure 2.3 [20].

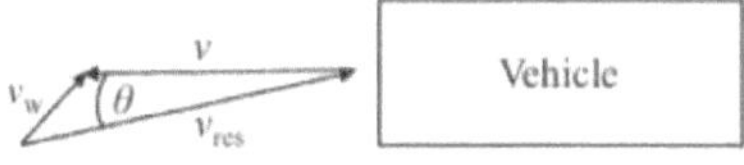

Figure 2.3: Definition of the sideslip angle (own representation based on [23]).

In [23] some typical curve progressions of the aerodynamic drag depending on the sideslip angle are shown. It can be seen, that with increasing sideslip angle the aerodynamic drag is increasing, too. A maximum value can be observed in the case the angle is large enough. At high values of θ (e.g. $\theta \gtrsim 25°$, as shown in [23]) even a decrease of the aerodynamic drag can be observed due to side forces which result in an aerodynamic push. However, these situations are rare (e.g. in gusts). Typical values of θ are less than $10°$ [23]. Furthermore, it can be assumed that the increase of the aerodynamic drag due to small sideslip angles θ is small [20].

<u>Ride height</u>
The influence o f t he r ide h eight c hange o n t he a erodynamic d rag w as i nvestigated by Schnepf et al. [24] and Schütz et al. [20]. In [24] on-road ride heights for different wheel designs on a BMW 3 Series sedan are compared to wind tunnel tests. The vehicle is elevated about 5 mm to 7 mm at a velocity of $140\,\mathrm{km/h}$. This ride height change increases the aerodynamic drag by 4 counts in the wind tunnel. And in [20], it is stated that a decrease in ride height of 30 mm at a velocity of $120\,\mathrm{km/h}$ results in a lower aerodynamic drag of about 20 counts for an AUDI Q7.

<u>c_w Dependency on the wind velocity</u>
From several wind tunnel studies, the effect of an aerodynamic drag coefficient reduction with increasing wind velocity is known, as shown in Figure 2.4. For the investigated vehicles (BMW Series 3 sedan, BMW Series 7 sedan and Rolls Royce Ghost) the aerodynamic drag coefficient at $80\,\mathrm{km/h}$ is 5 counts higher compared to the value at $140\,\mathrm{km/h}$.
In Weber's work [25], three different reasons for this effect are assumed:

- Changing of the flow field around the vehicle

- Wind tunnel interference

- Deformation of the vehicle surface

For further investigations, several numerical simulations with a BMW Series 3 sedan for wind velocities from $80\,\mathrm{km/h}$ to $160\,\mathrm{km/h}$ in the open jet were performed. The results show that changes in the flow fields (especially in the wake of the vehicle) and changes of the flow variables on the vehicle surface (like static pressure and wall shear stress) are the main reasons for this effect.

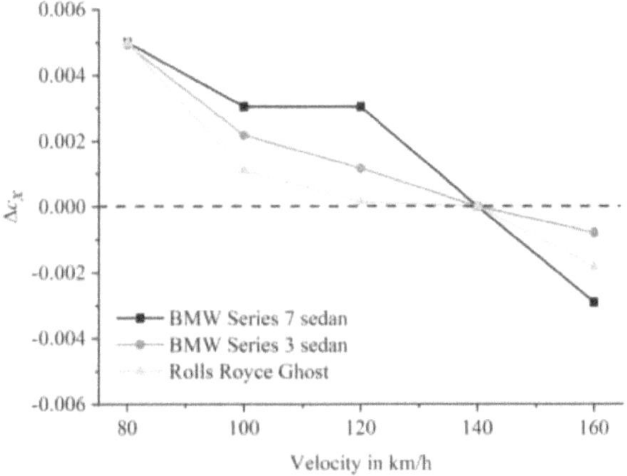

Figure 2.4: Difference between the aerodynamic drag measured at the velocity points of 80 km/h, 100 km/h, 120 km/h, 140 km/h and 160 km/h compared to the aerodynamic drag coefficient measured at 140 km/h for three different vehicles (adapted by permission from Sebastian Weber, Reynolds-Effekte in der Fahrzeugaerodynamik [25], 2018).

2.2.2 Rolling resistance

Due to the viscoelastic, hysteric properties of the tyre's rubber and structure, the tyre's elasticity can be described as a mechanical spring/damper system [17]. During the rotation of a tyre on a flat surface the tyre tread in the front area of the tyre contact area is compressed, whereby the force of the spring and damper is overcome. In contrast, at the rear area of the tyre contact area the tyre tread decompresses. At this point, only the spring force acts without the damper force, as the damper impedes the entire decompression of the spring. The result is an irregular surface pressure with the resulting force F_N, which acts with the distance of e_0 from the centre of the tyre, as shown in Figure 2.5, whereby, F_N is also the wheel load acting at the centre of the tyre. The moment of this force couple acts in the opposite direction to the wheel motion v_x. To overcome this moment a second force is necessary. This force, which acts in opposite direction of the motion direction, is denoted the rolling resistance F_{Roll} [16]:

$$F_{Roll} = \frac{e_0}{r_{stat}} \cdot F_N \qquad (2.14)$$

where:

F_{Roll}	is the rolling resistance in N;
e_0	is the distance between the centre of the wheel and the resulting force of the pressure surface of the tyre contact area in m;
r_{stat}	is the distance between wheel centre and road surface in m;
F_N	is the wheel load in N.

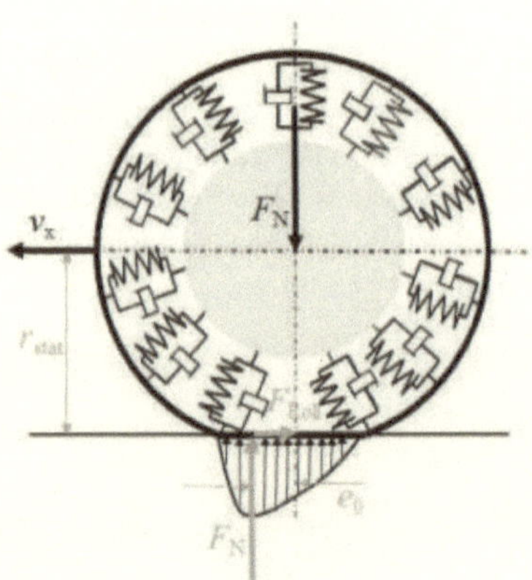

Figure 2.5: Tyre model for the explanation of the rolling resistance (adapted by permission from Hanser, Carl, Verlag GmbH & Co. KG, Grundlagen der Kraftfahrzeugtechnik [16], 2011).

Assuming that e_0 and r_{stat} are constant, the coefficient of rolling resistance f_r can be introduced [16, 17]:

$$f_{\text{r}} = \frac{F_{\text{Roll}}}{F_{\text{N}}} \tag{2.15}$$

where:

f_{r} is the coefficient of rolling resistance.

When the vehicle is driving on a dry surface without steering maneuvers, the total rolling resistance F_{Roll} is the sum of the following different components [17], which are also illustrated in Figure 2.6:

- Tyre rolling resistance $F_{\text{Roll,Tyre}}$

- Road rolling resistance $F_{\text{Roll,Road}}$

- Resistance due to tyre sideslip angle $F_{\text{Roll},\alpha}$

- Losses due to bearing friction and residual brake forces $F_{\text{Roll,Fric}}$

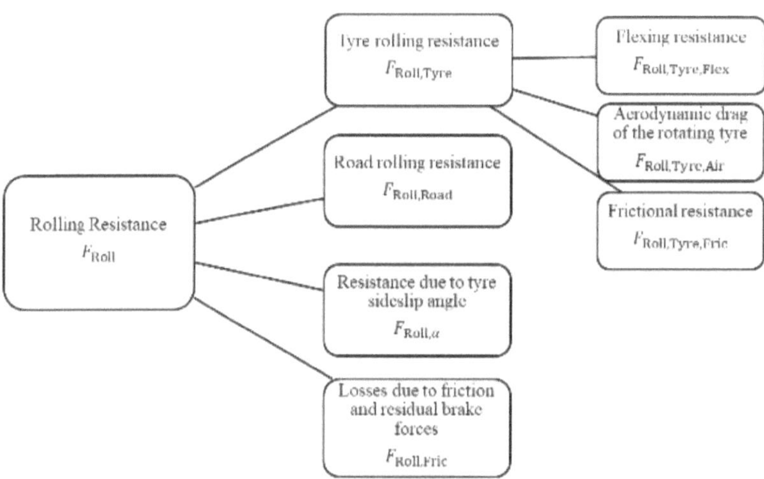

Figure 2.6: Composition of the rolling resistance of a vehicle F_{Roll}.

Tyre rolling resistance $F_{\text{Roll,Tyre}}$

The first component describes the resistance which has to be overcome to roll a tyre on an ideal, flat and dry surface without any steering maneuvers in the direction of travel. This resistance is characterized mainly by the chosen rubber compound and the construction of the tyre. It can be further separated into the components [17]:

- Flexing resistance $F_{\text{Roll,Tyre,Flex}}$

- Aerodynamic drag of the rotating tyre $F_{\text{Roll,Tyre,Air}}$

- Frictional resistance $F_{\text{Roll,Tyre,Fric}}$

The most important part is thereby the flexing resistance $F_{\text{Roll,Tyre,Flex}}$, which causes about 80 % to 95 % of the total rolling resistance F_{Roll}. Figure 2.7 illustrates a tyre's rolling motion with a certain wheel load F_{N} and an angular velocity ω. The resulting deformation of the tyre is described by the deformation of a spring/damper element (see also Figure 2.5) and the displacement s_T, whereby the energy of tyre deformation work that goes into elastic deformation is reversible during the tyre rebound. In contrast, the dampers' conversion of energy into heat is an energy loss and therefore irreversible. Thus, the flexing resistance $F_{\text{Roll,Tyre,Flex}}$ can be described by the following equation [17, 22, 26, 27, 28]:

$$F_{\text{Roll,Tyre,Flex}} = \frac{W_{\text{D,T,Flex}}}{s_{\text{T}}} \tag{2.16}$$

where:

$F_{\text{Roll,Tyre,Flex}}$	is the flexing resistance in N;
$W_{\text{D,T,Flex}}$	is the irreversible damping work in J;
s_{T}	is the displacement of the tyre deformation in m.

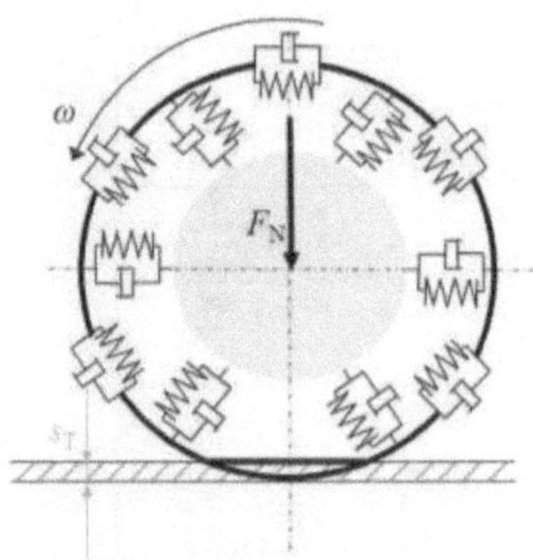

Figure 2.7: Flexing resistance $F_{\text{Roll,Tyre,Flex}}$ (adapted by permission from Springer Nature Customer Service Centre GmbH: Springer Nature, Chassis Handbook: Fundamentals, Driving Dynamics, Components, Mechatronics, Perspectives [17], 2011).

The second component of the tyre rolling resistance $F_{\text{Roll,Tyre}}$ describes the energy loss due to the air ventilation caused by the rotation of the wheel. This movement of air resists the wheel rotation motion and is called aerodynamic drag of the rotating tyre $F_{\text{Roll,Tyre,Air}}$. The drag increases with the square of the rotational velocity and is proportional to the aerodynamic resistance moment $M_{\text{Roll,Tyre,Air}}$ [17, 26, 28]. A further reason for this resistance is a surface friction and an unequal pressure distribution at the spokes of the rim. In Figure 2.8 an example for the unequal static pressure distribution is illustrated for a rotating vehicle tyre, which was investigated by a numerical simulation. It is shown, that in the rotational direction of the tyre the static pressure p_{stat} is higher in front of the spokes than behind [29]. This part of the aerodynamic drag of the rotating tyre $F_{\text{Roll,Tyre,Air}}$ is denoted wheel ventilation resistance F_{Vent}.

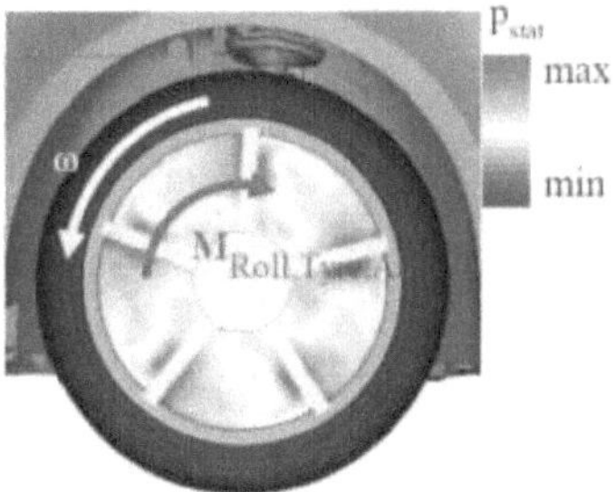

Figure 2.8: Numerical simulation of the pressure distributaion at a rotating vehicle tyre (adapted by permission from Springer Nature Customer Service Centre GmbH: Springer Nature, Analyse, Messung und Optimierung des aerodynamischen Ventilationswiderstands von Pkw-Rädern [29], 2018).

However, the losses due to the wheel ventilation are, similar to the other parts of the rolling resistance, internal forces in the wind tunnel and are usually not measured during the aerodynamic drag measurements in a wind tunnel with a 5-belt system. Nevertheless, an extended aerodynamic drag coefficient c_{w}^{*} can be defined as [20, 29]:

$$c_{\text{w}}^{*} = c_{\text{w}} + c_{\text{Vent}} \tag{2.17}$$

where:

c_{w}^{*}	is the extended aerodynamic drag coefficient;
c_{w}	is the aerodynamic drag coefficient;
c_{Vent}	is the aerodynamic drag coefficient of the rotating tyre or the wheel ventilation resistance coefficient, respectively.

The wheel ventilation resistance coefficient c_{Vent} is referenced to the vehicle frontal area A_x as the aerodynamic coefficient c_{w} [20, 29].

The last part $F_{\text{Roll,Tyre,Fric}}$ is caused by micro-slippage, which occurs during the rotating motion of the tyre between the tyre contact patch and the road surface. This leads to a relative motion between them. However, if the tyre is not braking or accelerating, this energy loss accounts for less than $5\,\%$ of the total rolling resistance F_{Roll}. This effect occurs not only between the tyre tread and the road surface, but also between the tyre and the wheel rim [17, 26, 30].

Road rolling resistance $F_{\text{Roll,Road}}$
The rolling resistance F_{Roll} also depends on the surface quality of the road. The resulting road rolling resistance $F_{\text{Roll,Road}}$ can be caused by an uneven road surface, a plastically deformable road surface and a resistance due to a water film on the road [17].

Resistance due to tyre sideslip angle $F_{\text{Roll},\alpha}$ caused by toe and camber
Until now, it is assumed that the central plane of the wheel is parallel to the driving direction of the vehicle. But in reality, the wheels are oriented at a toe angle δ_0 to the longitudinal axis of the vehicle, which also corresponds to a sideslip angle α. The resulting resistance due to the tyre toe angle $F_{\text{Roll,Toe}}$ is the sine term of the lateral tyre force F_{lat}, which is illustrated in Figure 2.9 and described by Equation 2.18 [17, 22].

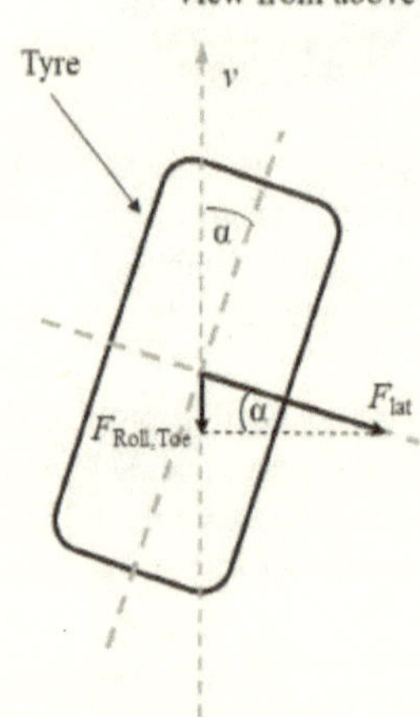

Figure 2.9: Illustration of the resistance $F_{\text{Roll,Toe}}$ due to a tyre toe angle δ_0, which is equal to the tyre sideslip angle α (own representation based on [16]).

$$F_{\text{Roll,Toe}} = F_{\text{lat}} \cdot \sin{(\alpha)} \tag{2.18}$$

where:

$F_{\text{Roll,Toe}}$	is the resistance due to the toe angle δ_0 in N;
F_{lat}	is the lateral tyre force in N;

α is the sideslip angle, which corresponds in this case to the toe angle δ_0 of the vehicle, in degrees.

For lateral acceleration from $3\,\mathrm{m/s^2}$ to $4\,\mathrm{m/s^2}$, which corresponds to toe angles from $1°$ to $2°$, the lateral tyre force F_{lat} is linearly proportional to the sideslip angle α, with the cornering stiffness C_α as a proportionality factor (see Figure 2.10) [31]. Then the lateral tyre force F_{lat} can be calculated as follows:

$$F_{\mathrm{lat}} = C_\alpha \cdot \alpha \tag{2.19}$$

where:

C_α is the cornering stiffness in $\mathrm{N/°}$.

For small angles α the expression $\sin(\alpha)$ can be approximated with α and the resulting resistance due to a tyre toe angle $F_{\mathrm{Roll,Toe}}$ can be estimated due to [16, 17, 32]:

$$F_{\mathrm{Roll,Toe}} = F_{\mathrm{lat}} \cdot \sin(\alpha) \approx F_{\mathrm{lat}} \cdot \alpha \approx C_\alpha \cdot \alpha^2 \tag{2.20}$$

Additionally, lateral tyre forces F_{lat} can be also caused by a wheel, which is inclined relative to the road surface in its longitudinal direction by the camber angle γ, as shown in Figure 2.11.
Similar to the lateral tyre force F_{lat} induced by the toe angle δ_0, the lateral tyre force F_{lat} induced by the camber angle γ can be estimated with the camber stiffness C_γ [22]:

$$F_{\mathrm{Roll,Camber}} = -C_\gamma \cdot \gamma \cdot \alpha \tag{2.21}$$

$F_{\mathrm{Roll,Camber}}$ is the resistance caused by camber in N;
C_γ is the camber stiffness in $\mathrm{N/°}$;
γ is the camber angle in degrees.

The total resistance caused by toe and camber $F_{\mathrm{Roll},\alpha}$ is defined as:

$$F_{\mathrm{Roll},\alpha} = F_{\mathrm{Roll,Toe}} + F_{\mathrm{Roll,Camber}} = (C_\alpha \cdot \alpha - C_\gamma \cdot \gamma) \cdot \alpha \tag{2.22}$$

where:

$F_{\mathrm{Roll},\alpha}$ is the total resistance caused by toe and camber in N.

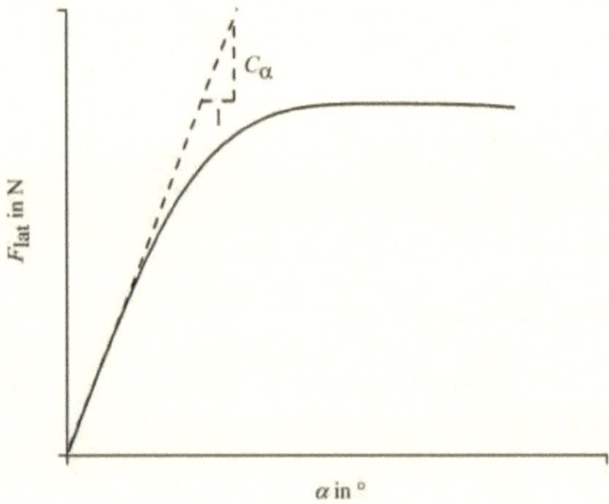

Figure 2.10: Lateral tyre force F_{lat} as function of the sideslip angle α (adapted by permission from Hanser, Carl, Verlag GmbH & Co. KG, Grundlagen der Kraftfahrzeugtechnik [16], 2011).

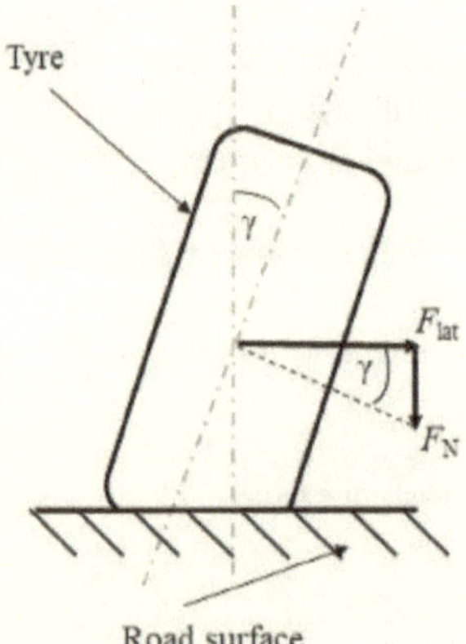

Figure 2.11: Illustration of the resulting lateral tyre forces F_{lat} due to a tyre camber angle γ (own representation based on [22]).

Losses due to friction and residual brake forces $F_{\text{Roll,Fric}}$

The last part of rolling resistance F_{Roll} is caused due to friction in the wheel bearing. However, this portion is very small compared to the rolling resistance of the tyre $F_{\text{Roll,Tyre}}$. Another portion is the residual brake force $F_{\text{Roll,Brake}}$, which is created by dragging brake pads on the brake disk. This resistance can also occur, when the brake pedal has already been released and the braking system is fully depressurized [17, 22]. During this off-brake phase, a friction contact can occur between the pad and the brake disk due to an unfavourable constellation of piston clearance and the dynamic runout. The piston clearance is defined as an axial distance between the brake pad and brake disk, when the brake is not applied [33].

Influencing factors on the rolling resistance

In the following, several factors influencing the rolling resistance are discussed. In order to provide a clear overview, the factors are divided into the three categories road/belt surface, experimental conditions and tyres, as shown in Figure 2.12.

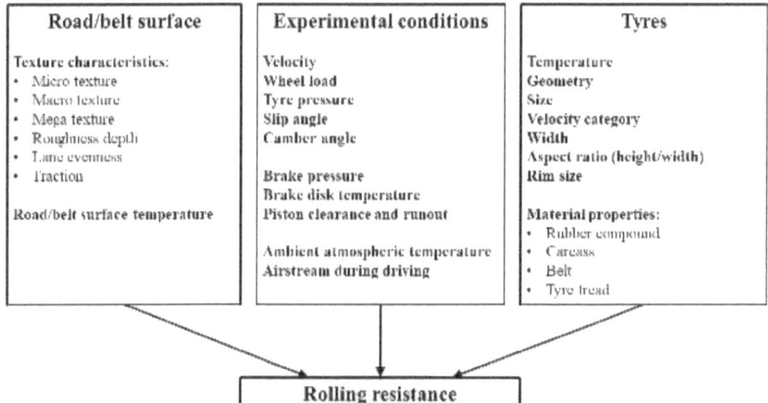

Figure 2.12: Overview of the most important influencing factors of the rolling resistance of a tyre divided into the categories road/belt surface, experimental conditions and tyres (adapted and extended by permission from Bundesanstalt für Straßenwesen, Vergleichsmessungen des Rollwiderstands auf der Straße und im Prüfstand [34], 1996).

Road/belt surface

The category road/belt surface contains the texture characteristics and the surface temperature of the road/belt surface, whereby the most important surface characteristics are roughness (macro and mega texture) and lane evenness. Unevenness of the road leads to wheel load fluctuations and tyre deformations. These cause hysteresis losses in the tyre rubber which result in higher energy losses. The wavelength of unevenness is defined in a range of 0.5 m to 50 m. In addition, while driving over a rough road surface further losses occur due to hysteresis of the tyre tread block deformation and due to mechanical interlocking of the tyre with the road surface. Therefore, the road surface can be described by the following texture characteristic with regard to the rolling resistance. The mega texture is defined by the manufacturing process of the road surface and has a wavelength of 50 mm to 500 mm. The macro roughness with a wavelength of 0.5 mm to 50 mm is affected by the composition of the surface material (for example of grain size distribution of the asphalt or concrete). The micro texture is defined by the material characteristics of the mineral substances (surfaces roughness), but has according to [35] no influence on rolling resistance. The wavelengths of this texture lie in a range between 0 mm and 0.5 mm [17, 35, 36].

According to this, Glaeser et al. found that the rolling resistance mostly increases proportionally for all investigated tyres with increasing grain diameter of the road surface. Thus, for the evaluation of the road texture influence on the rolling resistance, one single test tyre is sufficient for an investigation. It is found that in the case of an ideal road surface compared to a very uneven and rough road surface, the measured difference in rolling resistance is about 45 %[3] [36]. Moreover, Ullrich et al. [37] investigated the rolling resistance behaviour of four different tyres on eleven different road surfaces. Different absolute values were measured, but the influence of the texture was nearly independent of tyre type and size.

In Table 2.1 example rolling resistance coefficients for different road surfaces and a steel surface are listed [16][4]. Detailed information about the measurement procedure are not given. The values confirm the findings made by Glaeser et al. in [36]. Assuming the road surface is made of fine asphalt, the resulting rolling resistance can differ by about 20 %. Additionally, the difference in rolling resistance between a road surface with fine asphalt and coarse cement concrete can be up to 33.3 % [16].

Furthermore, in [38] the influence of different road surfaces and of Safety Walk[TM] on the rolling resistance coefficient of a vehicle tyre is investigated. The measurements are executed at an inner drum test bench at the Karlsruhe Institute of Technology. It was found that the rolling resistance of the same tyre for different wheel loads increases by about 20 % to 30 % on average on a real road surface compared to a Safety Walk[TM] surface. This effect is mainly explained by additional deformations caused by higher macro roughness

Table 2.1: Exemplary rolling resistance coefficients [16]

Road surface	Rolling resistance coefficients
Steel surface	0.008 ... 0.009
Fine asphalt	0.009 ... 0.011
Coarse cement concrete	0.010 ... 0.012
Tar	0.015 ... 0.020
Soil	0.030 ... 0.070
Field, sand, loose snow	0.100 ... 0.350

on real road surfaces.

Additionally, in Reimpell et al. [39] it is stated that the rolling resistance coefficient of Safety Walk^{TM} is about 5 % higher compared to the rolling resistance coefficient of steel. The coefficient for asphalt with a fine structure is about 21 % and for concrete with a rough surface structure even 40 % higher as compared to a steel surface. Detailed information about the measurement procedure is not given.

Furthermore, it is stated in [40] that a surface coated with Safety Walk^{TM} increases the tyre rolling resistance in a range from 2 % to 11 % as compared to a smooth steel surface. A description of the experimental conditions is also missing.

However, not only the surface texture but also the road surface temperature has an influence o n t he r olling r esistance. I n [19] a mongst o thers, t he i nfluence of th e road surface temperature on the tyre tread temperature and the rolling resistance coefficient is investigated. It is shown that the tyre tread temperature is directly influenced by the road surface temperature, as illustrated in Figure 2.13. It can be seen that with higher road surface temperature the tyre tread temperature also increases. On the other hand, the rolling resistance coefficient decreases. For a better overview, in Figure 2.14 the rolling resistance coefficient is plotted over the tyre tread temperature. It can be concluded that the rolling resistance increases by about 43 % with decreasing tyre tread temperature in the investigated tyre tread temperature range of 42 °C to 45.5 °C^5.

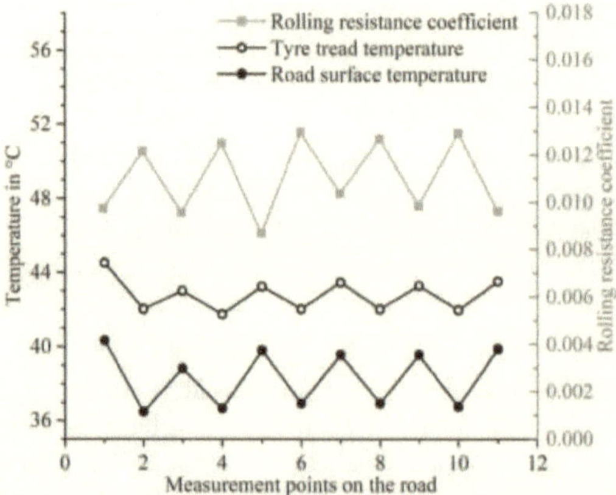

Figure 2.13: Dependence of the rolling resistance coefficient and the tyre tread temperature on the road surface temperature (adapted by permission from Narr Francke Attempto Verlag GmbH + Co. KG, Bestimmung und Aufteilung des Fahrwiderstandes im realen Fahrbetrieb [19], 2006).

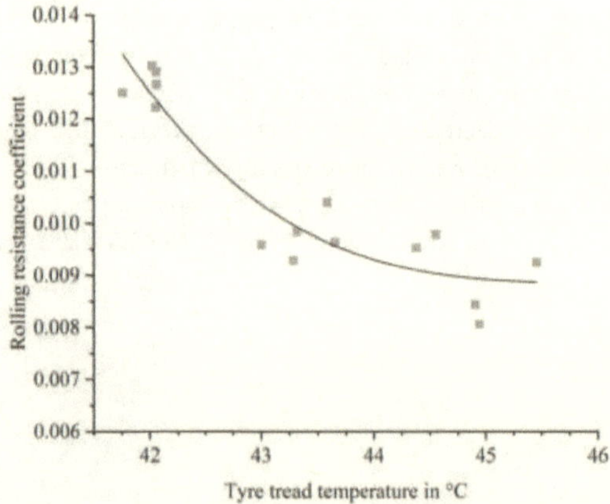

Figure 2.14: Dependence of the rolling resistance coefficient on the tyre tread temperature (adapted by permission from Narr Francke Attempto Verlag GmbH + Co. KG, Bestimmung und Aufteilung des Fahrwiderstandes im realen Fahrbetrieb [19], 2006).

Experimental conditions

The influence of the driving velocity on the rolling resistance is not linear. For the low and medium velocity range the rolling resistance is nearly constant. However, for high driving velocities the rolling resistance increases to the power of four with the driving velocity [16, 17, 22]. Figure 2.15 illustrates the almost constant and moderate increasing rolling resistance coefficient up to a velocity of about $100\,\mathrm{km/h}$ and the subsequent strong increase for radial and bias-ply tyres of passenger cars [17, 27].

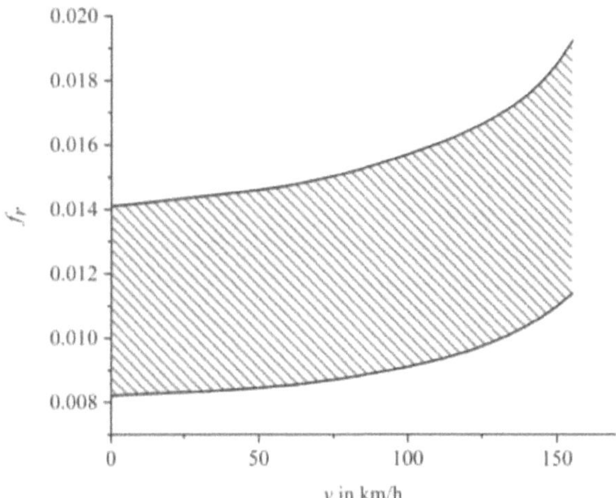

Figure 2.15: Rolling resistance coefficient f_r depending on the vehicle velocity v (adapted by permission from Springer Nature Customer Service Centre GmbH: Springer Nature, Vieweg Handbuch Kraftfahrzeugtechnik [27], 2013, and Chassis Handbook: Fundamentals, Driving Dynamics, Components, Mechatronics, Perspectives [17], 2011).

Consequently, the rolling resistance coefficient f_r can be estimated using the following equation [22]:

$$f_\mathrm{r} = f_\mathrm{r0} + f_\mathrm{r1} \cdot \left(\frac{v}{27.\overline{7}\,\mathrm{m/s}} \right) + f_\mathrm{r4} \cdot \left(\frac{v}{27.\overline{7}\,\mathrm{m/s}} \right)^4 \tag{2.23}$$

where:

f_r	is the rolling resistance coefficient;
f_r0	is the constant term of the rolling resistance coefficient f_r;
f_r1	is the first order term of the rolling resistance coefficient f_r;
v	is the vehicle velocity in $\mathrm{m/s}$;
f_r4	is the fourth order term of the rolling resistance coefficient f_r.

However, in Figure 2.15 a rolling resistance coefficient for a vehicle velocity of $0\,\mathrm{km/h}$ is given, although the rolling resistance has to be zero at this point. Regarding Equation 2.14, the rolling resistance is described by the distance between the centre of the wheel and the resulting force of the irregular pressure surface of the contact area, which is a result of a tyre rotation. In the case of a standing tyre, the distance e_0 and, therefore, also the rolling resistance are zero.

Furthermore, the rolling resistance increases degressively with increasing wheel load and decreases with increasing tyre's internal air pressure [17]. This relation can be estimated with the following equation, which is valid for passenger car tyres [26, 30]:

$$F_{\mathrm{Roll}} = F_{\mathrm{Roll,ISO}} \cdot \left(\frac{p_{\mathrm{Tyre}}}{p_{\mathrm{Tyre,ISO}}} \right)^{-0.4} \cdot \left(\frac{F_{\mathrm{N}}}{F_{\mathrm{N,ISO}}} \right)^{0.85} \tag{2.24}$$

where:

$F_{\mathrm{Roll,ISO}}$	is the rolling resistance measured according to ISO 8767 in N;
p_{Tyre}	is the tyre inflation air pressure in bar;
$p_{\mathrm{Tyre,ISO}}$	is tyre pressure as specified in ISO 8767 in bar;
$F_{\mathrm{N,ISO}}$	is the wheel load as specified in ISO 8767 in N.

As already mentioned previously, the toe and camber angles of the tyres have a significant influence on the rolling resistance. The resulting rolling resistance can be calculated according to Equation 2.22 using the cornering stiffness C_α and the camber stiffness C_γ. In Figure 2.16 the bandwidth of the cornering and camber stiffness for the tyre range of a vehicle is illustrated. It can be seen that the camber stiffness C_γ is only about $1/10$ of the cornering stiffness C_α. Therefore, the impact of camber angle γ on the rolling resistance is significantly lower than the impact of the toe angle δ_0 [31].
Typical values for toe angles δ_0 for a vehicle are [17]:

- Front axle of a vehicle with rear wheel drive: 0' to +30'

- Front axle of a vehicle with front wheel drive: -30' to +20'

- Rear axle: maximum -20' to +20'

Typical values for camber angles γ are [17]:

- -2° to +2°

According to Leister [30], the rolling resistance of a tyre increases by about $10\,\%$ due to a total toe angle of 1° (= 60'). In contrast, if the wheel has a camber angle of 1°, the rolling resistance increases only by $1\,\%$ to $2\,\%$. To validate this assumption, an example calculation is made. The following values are assumed:

- Wheel load F_{N}: $500\,\mathrm{kg} \cdot 9.81\,\mathrm{m/s^2} = 4905\,\mathrm{N}$

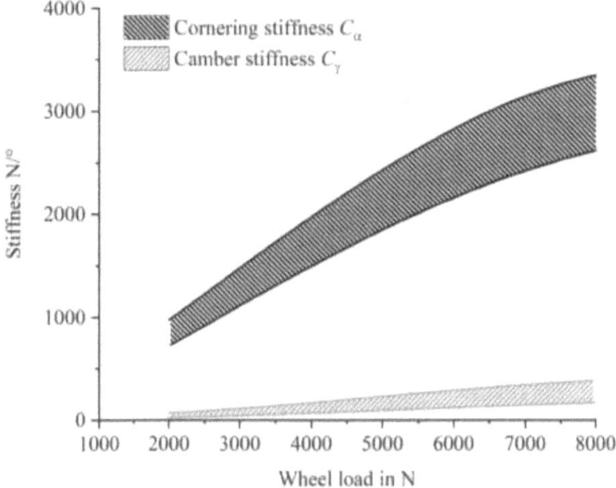

Figure 2.16: Bandwidth of cornering stiffness C_α and the camber stiffness C_γ for the tyre range of a vehicle (own representation based on [31]).

- Toe angle δ_0: 30'

- Maximum cornering stiffness for the given wheel load F_N: $C_\alpha = 2410\,\text{N/°}$ (see Figure 2.16)

- Camber angle γ: -1°

- Maximum camber stiffness for the given wheel load F_N: $C_\gamma = 220\,\text{N/°}$ (see Figure 2.16)

- Rolling resistance coefficient f_r for fine asphalt: 0.01 (see Table 2.1)

The resistance due to toe $F_{\text{Roll,Toe}}$ is calculated using Equation 2.20:

$$F_{\text{Roll,Toe}} \approx C_\alpha \cdot \alpha^2 = 2410\,\text{N/°} \cdot \frac{180°}{\pi} \cdot \left(\frac{30'}{60'/°} \cdot \frac{\pi}{180°} \right)^2 = 10.52\,\text{N} \tag{2.25}$$

Assuming a rolling resistance F_{Roll} of 49.05 N determined with Equation 2.15, the rolling resistance with an additional toe angle is increased by about 21.4 % and thus is significantly higher than the assumption made in [30].

The resulting resistance due to toe and an additional camber angle of -1° is determined with Equation 2.22:

$$F_{\text{Roll},\alpha} = (C_\alpha \cdot \alpha - C_\gamma \cdot \gamma) \cdot \alpha = 12.44\,\text{N} \tag{2.26}$$

Thus, the resistance due to an additional camber is increased by about $1.92\,\mathrm{N}$. This corresponds to an increase of the rolling resistance of $3.9\,\%$, which is also a higher value compared to the statement made in [30].

Additionally, the ambient atmospheric temperature and also the airstream during the driving have a direct influence on the tyre tread temperature and thus on the rolling resistance. In general, due to a higher ambient atmospheric temperature, the internal tyre temperature also increases, which leads to a lower amount of energy dissipated by the rubber components in the tyre when subjected to repeated deformation. As a consequence of the increased internal tyre temperature, the internal air pressure of the tyre also rises due to thermal expansion of the air in the tyre and therefore the tyre stiffness and deformations are reduced. As a result, the rolling resistance is lower at higher ambient atmospheric temperature [26, 40]. Furthermore, in [41], it is stated that due to an increase of the tyre core temperature of $10\,^\circ\mathrm{C}$, the cornering stiffness is decreased by about $3\,\%$ to $4\,\%$.

Therefore, to achieve a better comparability of the rolling resistance, which is measured at different atmospheric ambient temperatures, the rolling resistance F_{Roll} is adjusted to a reference temperature with a rolling resistance correction factor K_0 based on the following equation [10, 26, 40] :

$$F_{\mathrm{Roll,Ref}} = F_{\mathrm{Roll}} \cdot [1 + K_0 \cdot (T_{\mathrm{Amb}} - T_0)] \tag{2.27}$$

$F_{\mathrm{Roll,Ref}}$	is the rolling resistance of a tyre adjusted to reference temperature T_0 in N;
K_0	is the correction factor for the rolling resistance;
T_{Amb}	is the ambient atmospheric temperature in K;
T_0	is the reference temperature in K.

However, the rolling resistance is not a linear function of the ambient atmospheric temperature in the operation range of the tyre, as it is stated in [26], where the rolling resistance of a passenger car tyre F_{Roll} measured according to ISO 8767 is investigated depending on the ambient atmospheric temperature T_{Amb} [26].

Although the curve characteristic is not linear, it can be assumed that in the temperature range from $10\,^\circ\mathrm{C}$ to $40\,^\circ\mathrm{C}$ an ambient atmospheric temperature increase of $1\,^\circ\mathrm{C}$ corresponds to a reduction of the rolling resistance of about $0.6\,\%$ [26]. In [40] a reduction of rolling resistance of $0.5\,\%$ to even $0.8\,\%$ is regarded as being representative.

Ejsmont et al. [42] investigated amongst others the influence of the air temperature and different road surfaces on the rolling resistance of different tyres. The measurements are executed on the road and on test benches with different drum diameters ($1.7\,\mathrm{m}$, $2.0\,\mathrm{m}$ and $2.07\,\mathrm{m}$). They found that the magnitude of the air temperature influence on the resulting rolling resistance can differ up to $40\,\%$ for different tyres on the same drum surface. Moreover, in Figure 2.17 the influence of the air temperature between about $5\,^\circ\mathrm{C}$ and $35\,^\circ\mathrm{C}$ on the rolling resistance on different drum surfaces is shown. The investigated tyre is a mud and snow tyre of the manufacturer Avon (model AV4) with the size 195R14C. The used drum surfaces are described in Table 2.2. The test velocity is $80\,^{\mathrm{km}}\!/\!_{\mathrm{h}}$ [42].

Table 2.2: Drum surfaces used for the measurements [42]

Drum surface	Surface type	Test drum diameter
Drum surface 1	Replica of ISO reference surface	2.0 m
Drum surface 2	Poroelastic road surface	2.07 m
Drum surface 3	Poroelastic road surface	1.7 m
Drum surface 4	Replica of surface dressing 8/10 mm aggregate	1.7 m

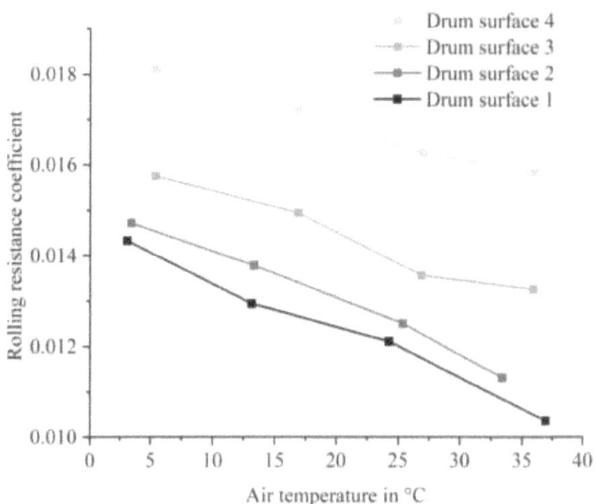

Figure 2.17: Influence of the air temperature on the rolling resistance coefficient for a mud and snow tyre of the manufacturerer Avon (model AV4) with the size 195R14C (adapted by permission from Springer Nature Customer Service Centre GmbH: Springer Nature, International Journal of Automotive Technology [42], 2017).

Similar to different tyres on the same drum surface, the magnitude of the air temperature influence on the resulting rolling resistance coefficients of the same tyre also depends on the chosen drum surface. The rolling resistance coefficient increases between about 6 % and 11 % depending on the chosen drum surface in the case that the air temperature is decreased from 23 °C to 10 °C. This corresponds to a reduction of the rolling resistance of between 0.5 % and 0.9 % due to an increase of the air temperature by 1 °C [42].

Furthermore, they showed that the tyre sidewall temperature is influenced by the pavement temperature, which is also affected by the air temperature, solar radiation and recent precipitations. In addition, the road temperature may not change as fast as the air temperature due to the higher thermal inertia [42, 43].

Another part of the rolling resistance are the residual brake forces. They depend amongst others on the brake pressure, brake disk temperature, piston clearance and runout of the brake disk. In the following, these influence factors are not investigated in detail. For more information about these influencing factors see in [33].

<u>Tyre</u>
Not only the road/belt surface and the experimental conditions, but also the tyre construction characteristics have a significant influence on the rolling resistance.

In particular, the losses which are caused due to deformation of the tyre $F_{\mathrm{Roll,Tyre,Flex}}$ are the highest part of the rolling resistance. They are characterized by the rubber compound, tyre tread profile and the construction of the tyre (carcass, belt, tyre tread). However, besides the construction details of the tyre also the rim size and rim design have an effect on the rolling resistance, especially on the wheel ventilation resistance F_{Vent} (see Figure 2.12).

To determine the influence of the rim size and design on the wheel ventilation resistance F_{Vent}, Link developed a test procedure in the wind tunnels with a 5-belt system of the BMW Group, FKFS (Forschungsinstitut für Kraftfahrwesen und Fahrzeugmotoren Stuttgart) and Porsche. Due to this procedure, the wheel ventilation resistance is measured with four load cells in x direction, which are connected to the four wheel drive units of the wind tunnel. However, according to this test procedure the wheel ventilation resistance is always measured as a part of the tyre rolling resistance F_{Roll}. To resolve the other parts of the rolling resistance, the wheel load is reduced by removing the vehicle springs. Additionally, the damper oil is drained. After this chassis modification the total weight of the vehicle is now supported by the rocker panel restraint system of the wind tunnel and the wheels stand only with their tare weight on the wheel drive units. For example for a front wheel, due to the reduction of the wheel load from 400 kg to 50 kg, the rolling resistance is reduced by about 90 %. As a result, at high velocities the wheel ventilation resistance is now the dominating part compared to the rolling resistance. At the beginning of the test procedure, there is a warm-up phase without incoming flow, which ensures comparable measurement conditions. Afterwards, two measurement phases with incoming flow at a high and a low velocity are following. The measurement phase at the low velocity of 40 km/h is used to determine the remaining losses due to friction and rolling. The measurements at the high velocity are executed at $180\,\mathrm{km/h}$. The resulting wheel ventilation resistance coefficient c_{Vent} can then be calculated with the subsequent equation:

$$c_{\mathrm{Vent}} = \frac{F_{\mathrm{Vent}}}{\frac{1}{2} \cdot \rho_0 \cdot A_x \left(v_{\mathrm{high}}^2 - v_{\mathrm{low}}^2 \right)} \tag{2.28}$$

with

$$F_{\mathrm{Vent}} = F_{x,v_{\mathrm{high}}} - F_{x,v_{\mathrm{low}}} \tag{2.29}$$

where:

F_{Vent}	is the wheel ventilation resistance in N;
$F_{x,v_{\mathrm{high}}}$	are the measured losses in x direction at the high velocity of $180\,\mathrm{km/h}$ in N;
$F_{x,v_{\mathrm{low}}}$	are the measured losses in x direction at the low velocity of $40\,\mathrm{km/h}$ in N;
ρ_0	is the dry air density, which is defined as $1.189\,\mathrm{kg/m^3}$;
A_x	is the frontal area of a vehicle in $\mathrm{m^2}$;
$v_{\mathrm{high/low}}$	is the wind and belt velocity at $180\,\mathrm{km/h}$ (high) and $40\,\mathrm{km/h}$ (low) using the test procedure for the determination of the wheel ventilation resistance in $\mathrm{m/s}$.

In Figure 2.18 five different rim setups are shown, which are investigated amongst others with regard to their individual wheel ventilation resistance. Setup 1 is a rim with 5 spokes. For the next two setups the spoke design is modified. At setup 2 the spoke corners are rounded and at setup 3 the spokes are concave. Setup 4 has the same spoke design as setup 1, but the number is increased up to 10 spokes. Moreover, the demonstrated results for the wheel ventilation resistance are based, inter alia, on the measurements executed in the BMW Group wind tunnel [29].

For the demonstrated cases, it can be seen that the spoke design has an influence of up to $2\,c_{\mathrm{Vent}}$ counts compared to the rim design with 5 spokes (setup 1). In contrast, the increase of the spoke numbers from 5 (setup 1) to 10 (setup 4) even results in an about $4\,c_{\mathrm{Vent}}$ counts higher wheel ventilation resistance coefficient [29].

As already explained, the most important step to use the test procedure developed by Link is the reduction of the rolling resistance. This is mainly realized by reducing the wheel load. Therefore, the vehicle springs are removed and the damper oil is drained. However, due to secondary spring rates resulting from the rubber bushings wheel load fluctuations during the measurements could occur which may cause measurement errors concerning the wheel ventilation resistance [6].

---

[6]BMW AG - Suspension, Damping, Roll Stabilization Non-Controlled, e-mail, 29.01.2021.

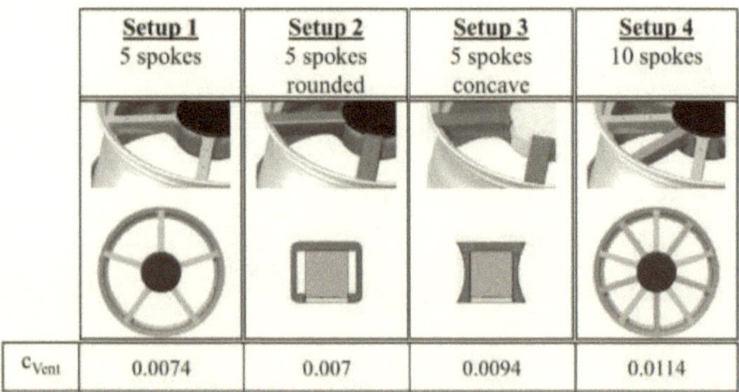

Figure 2.18: Wheel ventilation coefficients c_{Vent} for four different rim designs (own representation based on the results from [29] and figures adapted by permission from Springer Nature Customer Service Centre GmbH: Springer Nature, Analyse, Messung und Optimierung des aerodynamischen Ventilationswiderstands von Pkw-Rädern [29], 2018).

2.2.3 Drivetrain losses

The most important components of the vehicle powertrain are:

- Clutch

- Transmissions

- Drive shafts

- Axle drives

- Control systems, which act on the actuators and shift elements

In [44], a detailed overview of the basic influencing factors on the losses of mechanical transmissions is provided. The knowledge is mostly based on studies of Erler [45], Lauster [46], Leimann [47], Ohlendorf [48] and Goebbelet [49]. The most important findings taken from this overview are summarized in following.

Generally, the power losses of a gear transmission can be divided into the following parts [44, 46, 47, 48, 49] and [45] as cited in [44], which is also illustrated in Figure 2.19:

- Gearing losses P_{Gear}

- Bearing losses P_{Bearing}

- Seal losses P_{Seal}

- Splashing losses $P_{\text{Splashing}}$ (in case of splash lubrication)

- Losses due to further aggregates and components P_{Comp} (for example oil pump, switching components, air conditioning systems, ...)

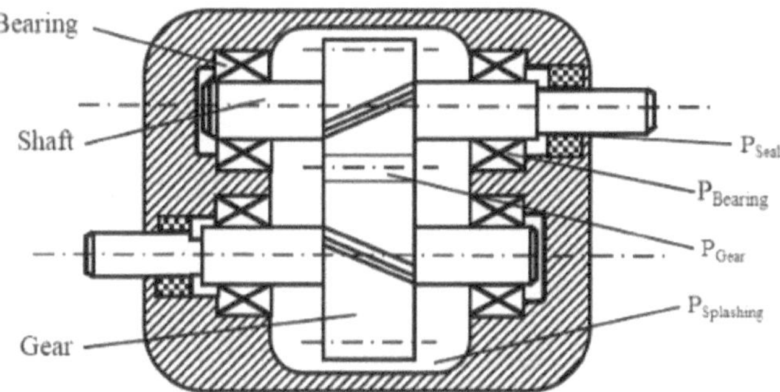

Figure 2.19: Power losses of a gear transmission (adapted by permission from FAT – Forschungsvereinigung Automobiltechnik e.V., Wirkungsgradmessung an Getrieben und Getriebeelementen [49], 1982).

Additionally, the losses are divided into load independent P_0 and load dependent losses P_{L}. The load independent losses P_0 consist of the gear losses, bearing losses, splashing losses and seal losses, whilst the gear transmission is running at no load. Otherwise, there are additional load dependent losses at the gears and bearings [44]. In [48] and as cited in [44] the parts of the power losses of a single stage spur gear with injection lubrication over the circumferential velocity and over the tooth load are investigated, exemplary.

It can be seen that the main part of the power losses is related to the gearing losses for a wide range. Only at low tooth loads are the bearing losses higher. Furthermore, with increasing circumferential velocity the load independent bearing losses increase; however, due to the decreasing gearing losses, the total losses for the gear transmission are also decreasing. This can be explained by the nearly constant gearing losses, but at the same time increasing power output with increasing circumferential velocity. However, in case of a gear transmission with splash lubrication, the gearing no-load losses would also increase with increasing circumferential velocity until a maximum value due to the additional splashing losses [44, 48].

In the next subsections, the single components of the drivetrain losses are described in detail. Afterwards, a short overview is provided, on how the drivetrain losses can be determined.

Gearing losses

In general, the load independent gearing losses occur due to fluid friction in the lubrication gap of the flanks, the oil displacement in the root clearence and both the acceleration and centrifuging of the oil in the case of injection lubrication. Some influencing factors on these losses are [44, 47, 48]:

- Oil viscosity

- Amount of oil in the contact area

- Location of the oil injection

- Circumferential velocity

- Whole depth

- Pressure angle

- Transverse contact ratio of the toothing

In [48] and as cited in [44] the load independent losses of a single stage spur gear are amongst others investigated depending on the circumferential velocity and the oil viscosity. For further information on the losses depending on the location of injection and on the amount of injected oil see in [48].

It is shown that the losses increase with the viscosity of the oil. Additionally, there is a maximum in the curve characteristics, which can be explained by the higher circumferential velocity. As a consequence, there has to be a faster displacement of the oil from the contact area and the root clearance, which results in higher losses. Simultaneously, the oil, which is injected over a certain time interval with increasing circumferential velocity, is distributed over a higher number of tooth spaces. Therefore, the amount of oil, which has to be displaced, is lower and the corresponding losses decrease [44, 48].

For the load dependent gearing losses the tribology conditions in the meshing have to be considered. The main influencing factors are [44, 46, 47, 48]:

- Roughness of the flanks

- Gear geometry

- Correction of toothing

- Tooth load

- Rotational velocity

- Temperature, density and viscosity of the oil

Bearing losses

The bearing losses for non-stationary operated gear transmission are mainly affected by rolling bearings [44, 46]. The load dependent part of the friction losses is mainly caused by elastic deformation of the rolling elements and bearing shells as well as by local sliding at the contact areas. The influencing factors are the amount of load, the type of bearing, the bearing dimension and the load direction. In contrast, the load independent part is dominated by the hydrodynamic losses of the lubricant. In this case, the influencing factors are rotational velocity, type of the bearing, the bearing reload, the type of the lubricant as well as the viscosity and the amount of lubricant [44, 47].

Splashing losses

In the case of splash lubrication, the lubricating effect is obtained due to the displacement and turbulence of the oil, when the gear immerses in the oil bath. The needed drive power is equal to the splashing losses. The losses are influenced by the following factors: Immersion depth, number of splashing gears, arrangement of the gears, direction of rotation, whole depth, rotational velocity, total amount of the oil, oil viscosity and construction of the gearbox housing [44, 48].

Furthermore, the oil temperature has a significant influence on the oil viscosity and therefore on the splashing losses. With increasing oil temperature the oil viscosity decreases [50], which results, besides the prior mentioned lower gearing losses (see in [48]), in lower splashing losses [51]. This relation is illustrated in Figure 2.20, where the temporal progression of the oil sump temperature and of the power losses are plotted. It can be seen that the power losses are decreasing with time due to the increasing oil sump temperature. Consequently, the higher oil temperature leads to lower splashing losses. However, the power losses as well as the oil temperature show the characteristics of a saturation curve. In the example displayed, the power losses can be reduced by about 33 % in the case that the saturation temperature of the oil in the sump is reached [51].

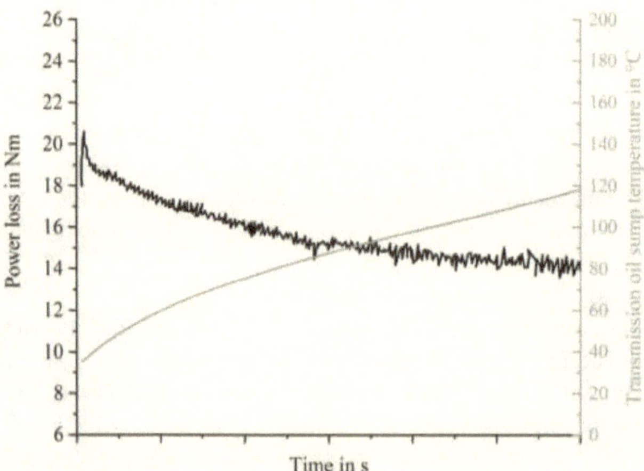

Figure 2.20: Power losses and corresponding oil sump temperature plotted over time (adapted by permission from Springer Nature Customer Service Centre GmbH: Springer Nature, The Automotive Transmission Book [51], 2015).

Furthermore, in [8] amongst others, the influence of higher oil temperatures of the rear axle differential and of the manual transmission of a BMW F33 430i Cabrio is investigated. For this purpose, the vehicle is driven by a flat belt dynamometer with a constant velocity for different time durations. This phase is called warm-up phase in the following. Afterwards, the measurement of the vehicle losses in x direction starts. It was found that due to a longer warm-up phase the oil temperatures in both transmission gears at the end of the warm-up phase increased by up to 5 °C. And as a consequence, the measured losses of the vehicle were reduced by about 5 N.

However, higher oil temperatures do not only reduce the splashing losses, but also friction losses, as it can be seen in Figure 2.21. In this figure the influence of two surface finishing (ground and super finished) of the gear flanks is plotted over the oil temperature. It shows that the friction coefficients and therefore also the friction losses decrease for both surface qualities with increasing oil temperature [51, 52].

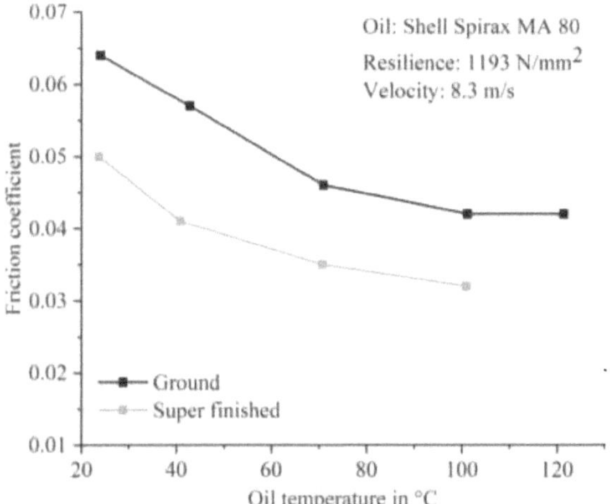

Figure 2.21: Influence of the surface finishing (ground super finished) and the oil temperature on the gear friction coefficient (adapted by permission from ZF Friedrichshafen AG, How to Minimize Power Losses in Transmissions, Axles and Steering Systems [52], 2010).

<u>Seal losses</u>
Influencing factors for the seal losses at the shafts are [44]:

- Lubrication

- Circumferential velocity of the running surface

- Seal diameter

- Material of the friction combination

- Treatment of the running surfaces

- Spring forces in radial direction

<u>Losses due to further aggregates and components</u>
Further losses occur due to shift elements and due to the oil pump regarding injection lubrication. The losses of the shift elements are affected mainly by the difference of rotational velocity of the shafts to be connected during the shift operation. These elements cause also losses in neutral position, if the elements are located behind the clutch viewed from the engine side as it can be seen in Figure 2.22 [44, 46, 51]. This figure illustrates the principal structure of an 8HP automatic transmission of ZF Friedrichshafen AG including a converter lockup clutch, a converter and four planetary gear sets.

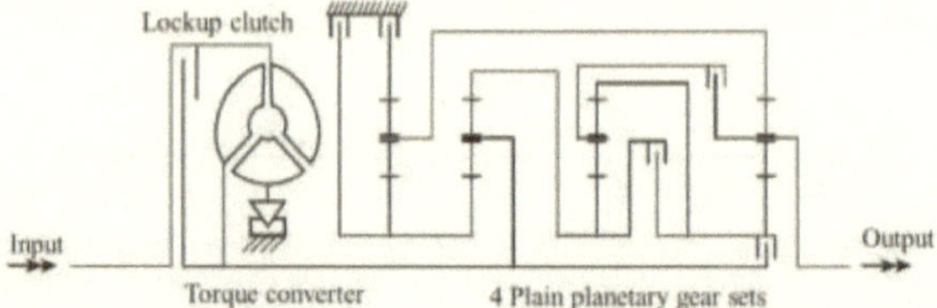

Figure 2.22: Principal structure of an 8HP automatic transmission of ZF Friedrichshafen AG (adapted by permission from Springer Nature Customer Service Centre GmbH: Springer Nature, The Automotive Transmission Book [51], 2015).

In Figure 2.23 the degrees of efficiency for the gears are shown by example for a 6HP and an 8HP automatic transmission [51]. In both cases the efficiency is generally decreasing with decreasing gear. A decreasing efficiency significates increasing drivetrain losses.
Furthermore, there are coupling losses, which consist mainly of fluid friction of wet-running multi-plate clutches and brakes in automatic transmissions and automated manual transmissions. Additional aggregates, as for example air conditioning systems, increase losses due to their power consumption [53]. These losses are only a component of the road load, determined by the methods introduced in subsection 2.3, if they are located behind the clutch (compare Figure 2.22).

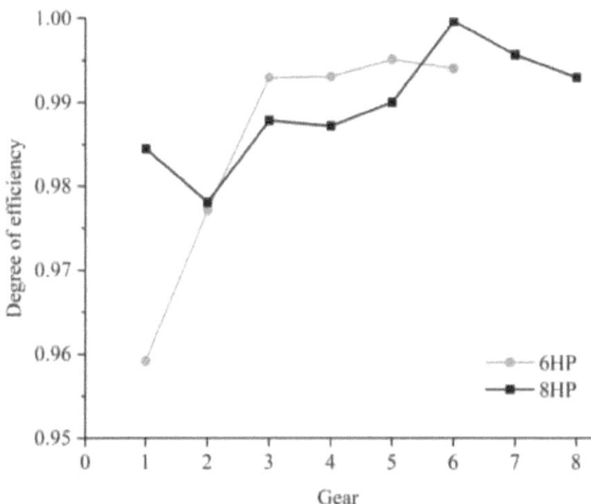

Figure 2.23: Degree of efficiency over the gears for a 6HP and an 8HP automatic transmission of ZF Friedrichshafen AG (adapted by permission from Springer Nature Customer Service Centre GmbH: Springer Nature, The Automotive Transmission Book [51], 2015).

Determination of drivetrain losses

For general power loss measurements of a gear transmissions, the following methods are proposed in [44, 49]:

- Back-to-back test rig

- Pendular suspension of the gear transmission

- Calorimetric measurement

- Pendulum machine

- Shaft torque meters

However, these measurement methods are only applicable when the gear transmission is to be investigated on its own. To measure the drivetrain losses of a vehicle, wheel torque transducers can be used, which replace the middle part of the rims [54]. The data transmission is usually carried out via a wireless digital telemetry system. A further method to determine the drivetrain losses of a vehicle is to place strain gauges on the output shafts of the vehicle [55]. Another method is to measure the drivetrain losses, for example at a Vehicle-in-the-Loop test bench (ViL-test bench), which generally enables, besides the determination of drivetrain losses, also the investigation, analysis and development of the drivetrain, chassis systems and energy management concepts [56]. Further examples for similar test benches are given in [57, 58].

In [26], it is stated that the internal frictional forces of the drivetrain for an average passenger car are around 50 N. However, it is mentioned that the internal friction is independent of the velocity, but includes additionally the losses due to the brake pads (compare subsection 2.2.2).

2.2.4 Climbing resistance

If a vehicle is driving uphill or downhill on a road with a slope angle α_S, a climbing resistance F_{Climb} acts in the longitudinal direction of the vehicle, which is equal to the sinusoidal component of the total vehicle weight force F_Z [17]:

$$F_{Climb} = F_Z \cdot \sin(\alpha_S) = m_{Veh} \cdot g_{Earth} \cdot \sin(\alpha_S) \tag{2.30}$$

where:

F_{Climb}	is the climbing resistance in N;
F_Z	is the total vehicle weight force in N;
α_S	is the slope angle of the road in degrees;
m_{Veh}	is the vehicle mass in kg;
g_{Earth}	is the earth's gravity in $\mathrm{m/s^2}$.

The climbing resistance F_{Climb} is a conservative force. Hence, the required energy by the vehicle's engine to overcome this resistance is saved as potential energy and can be recovered [17].

2.2.5 Inertial resistance

In contrast to a stationary drive with constant velocity, during an unsteady drive an additional inertial resistance F_{Inertial} acts due to accelerating and braking manoeuvres. In the case of accelerating, the inertial resistance F_{Inertial} has to be overcome by the vehicle's powertrain. This resistance can be divided into a translational and a rotatory component. The translational portion F_{Trans} means the unsteady motion of the vehicle mass and the rotatory portion F_{Rot} results due to the acceleration and deceleration of vehicle components. Rotating masses are gear, cardan and drive shafts, gear wheels, axle drives and the vehicle wheels. The inertial resistance can be calculated with the following equation:

$$F_{\text{Inertial}} = F_{\text{Trans}} + F_{\text{Rot}} = m_{\text{Veh}} \cdot a + \frac{J_{\text{red}}}{r_{\text{dyn}}} \cdot \dot{\omega} = e_{\text{mass}} \cdot m_{\text{Veh}} \cdot a \qquad (2.31)$$

where:

F_{Inertial}	is the inertial resistance force in N;
F_{Trans}	is the translational force in N;
F_{Rot}	is the rotational force in N;
m_{Veh}	is the vehicle mass in kg;
a	is the translational acceleration in m/s^2;
J_{red}	is the reduced rotational inertia of the vehicle components in kg $\cdot$ m^2;
r_{dyn}	is the dynamic rolling radius in m;
$\dot{\omega}$	is the rotational acceleration in 1/s^2;
e_{mass}	is the mass factor.

Comparable to the climbing resistance F_{Climb}, the inertial resistance F_{Inertial} is a conservative force and the energy needed to accelerate the vehicle and its rotating masses is converted into kinetic energy [16, 17, 18, 22]. In Table 2.3 values of $(e_{\text{mass}} - 1)$[7] for a 5-speed transmission are given [16].

---

[7] In [16] these values are defined as the mass factor e_{mass}, erroneously.

Table 2.3: Values of $(e_{mass} - 1)$ for a 5-speed transmission [16]

Gear	$(e_{mass} - 1)$
1	0.25 ... 0.50
2	0.11 ... 0.21
3	0.06 ... 0.11
4	0.4 ... 0.08
5	0.04 ... 0.06
Neutral	0.03 ... 0.05

2.3 Methods of road load determination

Since inception of WLTP there are also other methods allowed besides the road load determination on a proving ground. In the following subsections the different methods to determine the road load of a vehicle are described in detail.

2.3.1 Road load determination using the coastdown method on a proving ground

In general, the road load of a vehicle can be determined on a proving ground. Here two options are allowed according to GTR No. 15 [10]:

- Coastdown method with stationary anemometer

- Coastdown method with on-board anemometer

In the following, only the test procedure with stationary anemometer is explained. Further information about the differences in the measurement procedure and in the correction equations for the test procedure with on-board anemometer is given in [10].
The proving ground for the CoastDown Method (CDM) consists usually of an oval test track with an obligatory driving direction A and B, as shown schematically in Figure 2.24. If an oval proving ground is chosen, two stationary anemometers are necessary to measure the wind velocity and wind direction at each part of the test track. The stationary anemometers, the two slip roads onto the test track and the test track exit are also illustrated in this figure.
Additionally, there are several requirements placed on weather and test track conditions, which have to be considered during the execution of the road load measurement [10]:

- The measurements using a stationary anemometer are valid, when the wind velocity is less than $5\,\mathrm{m/s}$ over a period of 5 seconds on average. Additionally, the vector component of the wind velocity across the roadway of the proving ground has to be less than $2\,\mathrm{m/s}$. (For information about the permissible wind conditions for using on-boad anomemeter see [10])

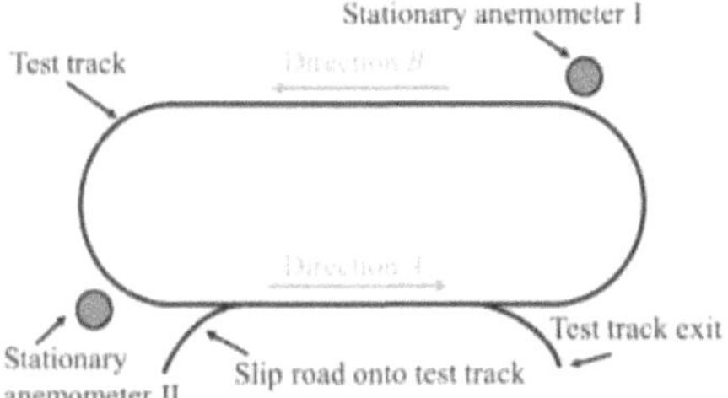

Figure 2.24: Proving ground with an oval test track and obligatory driving direction A and B, slip road onto the test track, test track exit and stationary anemometers.

- For a valid test, the atmospheric temperature has to be in a range between $5\,°C$ and $40\,°C$.

- The road surface of the test track has to be flat, even, clean and dry. In addition, it has to be free of wind barriers and obstacles, which might have an influence on the measurement. Moreover, the texture and composition of the road surface has to represent current urban and highway road surfaces. The maximum sum of longitudinal slopes of the parallel test track segments of the road on an oval test track has to be in a range between $0\,\%$ and $0.1\,\%$ with a maximal camber of not more than $1.5\,\%$.

In the following, the test procedure to determine the road load on a proving ground is described. Afterwards, the calculation, including the correction terms to standard reference conditions, is explained.

Before the measurement procedure starts, the vehicle accelerates to $80\,\mathrm{km/h}$ and then decelerates to $20\,\mathrm{km/h}$ within 5 to 10 seconds by moderate braking and with the clutch disengaged or automatic transmission placed in neutral. This braking phase shall ensure that the determined road load contains residual brake forces. During the subsequent test procedure, there is no further actuating of the braking system allowed. In the next step, the vehicle is driven at $90\,\%$ of the maximum velocity of the applicable WLTC for at least $1200\,\mathrm{s}$, until stable conditions are reached. For example, for vehicles with a maximum velocity of $\geq 120\,\mathrm{km/h}$ (WLTC Class 3b) the warm-up velocity is $118\,\mathrm{km/h}$. After the warm-up phase, the vehicle accelerates $10\,\mathrm{km/h}$ to $15\,\mathrm{km/h}$ above the highest reference velocity point v_j of $130\,\mathrm{km/h}$. The measurement phase (coastdown run) starts immediately afterwards. During the measurement phase, the transmission is placed in neutral and any steering movement or vehicle brake system actuation have to be avoided as much as possible. During the coastdown runs, the time corresponding to the reference velocity v_j as the elapsed time from the vehicle velocity $(v_j + 5\,\mathrm{km/h})$ to $(v_j - 5\,\mathrm{km/h})$ is measured (t_{jAi} and t_{jBi}). The reference velocity points v_j are from $130\,\mathrm{km/h}$ to $20\,\mathrm{km/h}$ in incremental steps of $10\,\mathrm{km/h}$. Moreover, coastdown runs are carried out in opposite directions (direction A and direction B, as shown in Figure 2.24), to eliminate influences of the wind (as wind direction and wind velocity) and of the test track (like slope in longitudinal direction), until a minimum of three pairs of measurements have been obtained, which satisfy the statistical precision p_j. The statistical precision p_j is defined as [10]:

$$p_j = \frac{h \cdot \sigma_j}{\sqrt{n} \cdot \Delta t_{\mathrm{p}j}} \leq 0.03 \tag{2.32}$$

with

$$\Delta t_{\mathrm{p}j} = \frac{n}{\sum\limits_{i=1}^{n} \frac{1}{\Delta t_{ji}}}, \tag{2.33}$$

$$\Delta t_{ji} = \frac{2}{\left(\frac{1}{t_{jAi}}\right) + \left(\frac{1}{t_{jBi}}\right)} \tag{2.34}$$

and

$$\sigma_j = \sqrt{\frac{1}{n-1} \sum_{i=1}^{n} (\Delta t_{ji} - \Delta t_{\mathrm{p}j})^2} \tag{2.35}$$

where:

p_j	is the statistical precision of the measurements determined at the reference velocity v_j;
j	is the counter variable for the reference velocity point v_j (20 km/h, 30 km/h, ..., 130 km/h);
h	is a coefficient due to the Student t-distribution (see Table 2.4);
σ_j	is the standard deviation of the coastdown time in s (see Equation 2.35);
n	is the number of pairs of measurements/coastdown runs;
$\Delta t_{\mathrm{p}j}$	is the harmonic average of the coastdown time at the reference velocity v_j in s (see Equation 2.33);
Δt_{ji}	is the harmonic average coastdown time of the ith pair of the measurements at the reference velocity point v_j in s (see Equation 2.34);
t_{jAi}	is the coastdown time of the ith measurement at the reference velocity point v_j in direction A in s;
t_{jBi}	is the coastdown time of the ith measurement at the reference velocity point v_j in direction B in s.

The Student t-distribution is used, if there is only a small number of samples n (In this case it corresponds to the number of pairs of measurements.) [13]. In Table 2.4 some coefficients h due to the Student t-distribution considering a two-sided confidence interval of 95 % are listed [59]. These values are also used according to GTR No. 15 [10].

However, the straight segments of the test track are usually not long enough to carry out each coastdown run from 135 km/h to 15 km/h without interruption. Thus, split runs are performed. The split runs can be carried out for example from 135 km/h to 75 km/h and from 75 km/h to 15 km/h. The above described test procedure for the road load determination

Table 2.4: Coefficients due to the Student t-distribution h considering a two-sided confidence interval of 95 % [59]

Number of samples n	Degree of freedom	h
3	2	4.303
4	3	3.182
5	4	2.776
6	5	2.571
...	...	...

using the coastdown method on a proving ground is illustrated schematically in Figure 2.25.

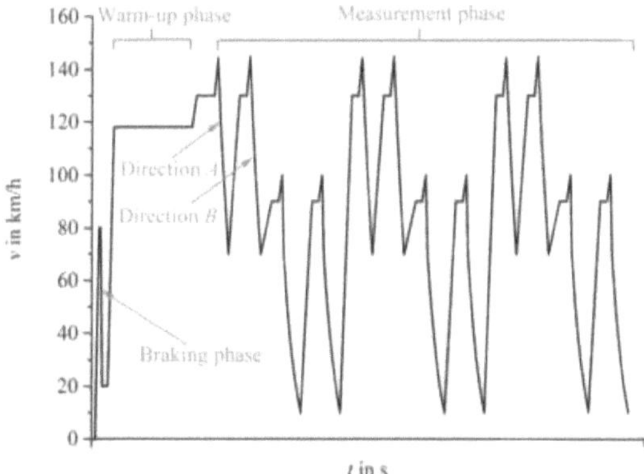

Figure 2.25: Schematic representation of the coastdown method on a proving ground with braking phase, warm-up phase and the measurement phase, where single coastdown runs are executed in the driving direction A and B and with split runs (adapted by permission from Isabell Vogeler: Different methods for road load determination in comparison: Wind tunnel, Wind tunnel method according to WLTP and Coastdown method [60], 2018).

Afterwards, the uncorrected road load of the vehicle F_j can be calculated with the following equations [10] [8]:

$$F_j = (m_{\mathrm{av}} + m_{\mathrm{r}}) \cdot a_j \tag{2.36}$$

with

---

[8]In [10], an additional factor of 2 is stated in Equation 2.36, erroneously.

$$a_j = \frac{\Delta v}{\Delta t_j}, \tag{2.37}$$

$$m_{av} = \frac{m_1 + m_2}{2}, \tag{2.38}$$

and

$$\Delta v = (v_j + 1.3\overline{8}\,\text{m/s}) - (v_{j+1} + 1.3\overline{8}\,\text{m/s}) = 2.\overline{8}\,\text{m/s} \, (= 10\,\text{km/h}) \tag{2.39}$$

where:

F_j is the uncorrected road load of the vehicle at the reference velocity point v_j in N (see Equation 2.36);

m_{av} is the arithmetic average of the test vehicle masses at the beginning m_1 and at the end m_2 of the coastdown test procedure in kg (see Equation 2.38);

m_r is the the equivalent effective mass of rotating components (for more information see in [10]) in kg (see Equation 2.40);

a_j is the deceleration at reference velocity point v_j in m/s^2 (see Equation 2.37);

Δv is the incremental step between two reference velocity points ($= 10\,\text{km/h}$) in m/s (see Equation 2.39);

Δt_j is the harmonic average of the alternate coastdown time measurements in direction A and B at the reference velocity point v_j in s (see Equation 2.42);

m_1 is the test vehicle mass at the beginning of the coastdown test procedure in kg;

m_2 is the test vehicle mass at the end of the coastdown test procedure in kg;

v_j, v_{j+1} is the vehicle velocity at the reference velocity point j in m/s.

Using Equation 2.37, it is assumed that the deceleration between the two velocity points $(v_j + 1.3\overline{8}\,\text{m/s})$ and $(v_j - 1.3\overline{8}\,\text{m/s})$ is constant. In combination with the corresponding averaged elapsed time Δt_j (see Equation 2.42) between these two velocity points, the deceleration a_j can be calculated with Equation 2.37. Due to this assumption, the complex integration steps described in section 2.1 can be avoided. The validity of this simplification is shown in section A.5 in the appendix.

Furthermore, to compensate for the rotating masses, which reduce the road load of a vehicle during the coastdown run, a correction using the equivalent effective mass m_r is inserted. The equivalent effective mass m_r is defined as the mass of all wheels and vehicle components, which are rotating while the gearbox is placed in neutral. It can be calculated as follows [10]:

$$m_r = (e_{\text{mass}} - 1) \cdot (m_{\text{MIRO}} + 25\,\text{kg}) \tag{2.40}$$

with

$$(e_{\text{mass}} - 1) = 0.03 \tag{2.41}$$

where:

m_{MIRO} is the sum of the mass in running order in kg;

e_{mass} is the mass factor (see Equation 2.41).

The Mass In Running Order (MIRO) m_{MIRO} is defined as the vehicle mass fitted with the standard equipment in accordance with the manufacturer's specifications including the following requirements [61]:

- Fuel tank filled to at least 90 % of its capacity

- Mass of the driver (75 kg)

- Mass of the fuel and liquids

- Mass of the bodywork, additional couplings and spare wheels (In the case that they are fitted.)

The factor $(e_{\text{mass}} - 1)$ considers the rotating masses and has in this case a value of 0.03 [10]. This value is in the same range as the values for a 5-speed transmission with the gear in neutral (compare Table 2.3).

The harmonic average of the alternate coastdown time measurements in direction A and B is calculated using the following equation [10]:

$$\Delta t_j = \frac{2}{\frac{1}{\Delta t_{jA}} + \frac{1}{\Delta t_{jB}}} \tag{2.42}$$

whereby the harmonic average coastdown times in the directions A and B are defined as:

$$\Delta t_{jA} = \frac{n}{\sum\limits_{i=1}^{n} \frac{1}{t_{jAi}}} \tag{2.43}$$

and

$$\Delta t_{jB} = \frac{n}{\sum\limits_{i=1}^{n} \frac{1}{t_{jBi}}} \tag{2.44}$$

where:

Δt_{jA} is the harmonic average of the coastdown times in direction A at the reference velocity point v_j in s (see Equation 2.43);

Δt_{jB} is the harmonic average of the coastdown times in direction B at the reference velocity point v_j in s (see Equation 2.44).

The road load coefficients f_0, f_1 and f_2 can then be calculated with a polynomial regression of second order (see Equation 2.2). However, not all road load measurements are executed at the same weather conditions and, therefore, the determined road loads are not comparable. Thus, several correction factors are defined, which have to be applied to the determined road load coefficients to achieve the same reference atmospheric conditions [10]:

- Air resistance correction factor K_2

- Rolling resistance correction factor K_0 (The name is according to [10]. However, the factor is applied to the road load coefficients f_0 and f_1, which do not only contain the rolling resistance part but also the drivetrain losses, see section 2.5.)

- Wind correction term w_1

- Test mass correction factor K_1

The reference atmospheric conditions according to WLTP are defined as [10]:

- Atmospheric pressure: $p_0 = 100\,\text{kPa}$

- Atmospheric temperature: $T_0 = 20\,°\text{C}$

- Dry air density: $\rho_0 = 1.189\,\text{kg/m}^3$

- Wind velocity: $0\,\text{m/s}$

With the aid of these correction factors the road load $F_{j,\text{CDM}}$ of a vehicle corrected to reference conditions can be calculated with the following equations [10]:

$$F_{j,\text{CDM}} = ((f_0 - w_1 - K_1) + f_1 \cdot v_j) \cdot \left(1 + K_0 \cdot \left(\overline{T}_\text{Amb} - T_0\right)\right) + K_2 \cdot f_2 \cdot v_j^2 \tag{2.45}$$

with

$$K_0 = 8.6 \cdot 10^{-3}\,\text{K}^{-1} \tag{2.46}$$

$$K_1 = f_0 \cdot \left(1 - \frac{m_\text{test}}{m_\text{av}}\right) = f_0 \cdot \left(1 - \frac{2 \cdot m_\text{test}}{m_1 + m_2}\right) \tag{2.47}$$

$$K_2 = \frac{\overline{T}_\text{Amb}}{293\,\text{K}} \cdot \frac{100\,\text{kPa}}{\overline{p}_\text{Amb}} \tag{2.48}$$

and

$$w_1 = f_2 \cdot \overline{v}_\text{Wind}^2 \tag{2.49}$$

where:

$$(e_{\mathrm{mass}} - 1) = 0.03 \tag{2.41}$$

where:

m_{MIRO} is the sum of the mass in running order in kg;

e_{mass} is the mass factor (see Equation 2.41).

The Mass In Running Order (MIRO) m_{MIRO} is defined as the vehicle mass fitted with the standard equipment in accordance with the manufacturer's specifications including the following requirements [61]:

- Fuel tank filled to at least 90 % of its capacity

- Mass of the driver (75 kg)

- Mass of the fuel and liquids

- Mass of the bodywork, additional couplings and spare wheels (In the case that they are fitted.)

The factor $(e_{\mathrm{mass}} - 1)$ considers the rotating masses and has in this case a value of 0.03 [10]. This value is in the same range as the values for a 5-speed transmission with the gear in neutral (compare Table 2.3).

The harmonic average of the alternate coastdown time measurements in direction A and B is calculated using the following equation [10]:

$$\Delta t_j = \frac{2}{\frac{1}{\Delta t_{jA}} + \frac{1}{\Delta t_{jB}}} \tag{2.42}$$

whereby the harmonic average coastdown times in the directions A and B are defined as:

$$\Delta t_{jA} = \frac{n}{\sum_{i=1}^{n} \frac{1}{t_{jAi}}} \tag{2.43}$$

and

$$\Delta t_{jB} = \frac{n}{\sum_{i=1}^{n} \frac{1}{t_{jBi}}} \tag{2.44}$$

where:

Δt_{jA} is the harmonic average of the coastdown times in direction A at the reference velocity point v_j in s (see Equation 2.43);

Δt_{jB} is the harmonic average of the coastdown times in direction B at the reference velocity point v_j in s (see Equation 2.44).

The road load coefficients f_0, f_1 and f_2 can then be calculated with a polynomial regression of second order (see Equation 2.2). However, not all road load measurements are executed at the same weather conditions and, therefore, the determined road loads are not comparable. Thus, several correction factors are defined, which have to be applied to the determined road load coefficients to achieve the same reference atmospheric conditions [10]:

- Air resistance correction factor K_2

- Rolling resistance correction factor K_0 (The name is according to [10]. However, the factor is applied to the road load coefficients f_0 and f_1, which do not only contain the rolling resistance part but also the drivetrain losses, see section 2.5.)

- Wind correction term w_1

- Test mass correction factor K_1

The reference atmospheric conditions according to WLTP are defined as [10]:

- Atmospheric pressure: $p_0 = 100\,\mathrm{kPa}$

- Atmospheric temperature: $T_0 = 20\,^\circ\mathrm{C}$

- Dry air density: $\rho_0 = 1.189\,\mathrm{kg/m^3}$

- Wind velocity: $0\,\mathrm{m/s}$

With the aid of these correction factors the road load $F_{j,\mathrm{CDM}}$ of a vehicle corrected to reference conditions can be calculated with the following equations [10]:

$$F_{j,\mathrm{CDM}} = ((f_0 - w_1 - K_1) + f_1 \cdot v_j) \cdot \left(1 + K_0 \cdot \left(\overline{T}_{\mathrm{Amb}} - T_0\right)\right) + K_2 \cdot f_2 \cdot v_j^2 \qquad (2.45)$$

with

$$K_0 = 8.6 \cdot 10^{-3}\,\mathrm{K^{-1}} \qquad (2.46)$$

$$K_1 = f_0 \cdot \left(1 - \frac{m_{\mathrm{test}}}{m_{\mathrm{av}}}\right) = f_0 \cdot \left(1 - \frac{2 \cdot m_{\mathrm{test}}}{m_1 + m_2}\right) \qquad (2.47)$$

$$K_2 = \frac{\overline{T}_{\mathrm{Amb}}}{293\,\mathrm{K}} \cdot \frac{100\,\mathrm{kPa}}{\overline{p}_{\mathrm{Amb}}} \qquad (2.48)$$

and

$$w_1 = f_2 \cdot \overline{v}_{\mathrm{Wind}}^2 \qquad (2.49)$$

where:

$F_{j,\text{CDM}}$	is the road load of the vehicle corrected to reference conditions at the reference velocity point v_j using the coastdown method (CDM) in N;
f_0	is the constant term of the road load coefficients in N;
f_1	is the coefficient of the first order term in $\text{N}/(\text{m/s})$;
f_2	is the coefficient of the second order term in $\text{N}/(\text{m/s})^2$;
K_0	is the correction factor for rolling resistance (and drivetrain losses) due to difference in air temperature compared to the reference condition at $20\,^{\circ}\text{C}$ in K^{-1};
K_1	is the test mass correction factor, which considers the test mass differences of the vehicle between the averaged test mass m_{av} and the reference test mass m_{Veh}, in N;
m_{test}	is the vehicle test mass defined for the road load determination in kg;
K_2	is the correction factor for air resistance due to differences in air pressure and temperature compared to the reference conditions in N;
$\overline{T}_{\text{Amb}}$	is the arithmetic average ambient atmospheric temperature in K;
$\overline{p}_{\text{Amb}}$	is the arithmetic average pressure in kPa;
T_0	is the reference atmospheric temperature of $20\,^{\circ}\text{C}$ in K;
w_1	is the wind resistance due to wind in opposite directions alongside the road of the test track during the measurement procedure in N;
$\overline{v}_{\text{Wind}}$	is the arithmetic average wind velocity alongside the test track during the measurement phase, which cannot be cancelled out by alternate coastdown runs in direction A and B, in m/s.

The composition of Equation 2.45 is explained in the following:

- The vehicle mass corresponds to a constant wheel load. Therefore, the test mass correction factor K_0, which considers the difference between the prior defined vehicle test mass m_{test} and the real averaged vehicle mass m_{av} during the test procedure, is only applied to the constant term f_0 of the road load coefficients.

- Using the coastdown method according to GTR No. 15 the wind alongside the road of the test track is corrected by subtracting the wind difference $\overline{v}_{\text{Wind}}$, which cannot be cancelled out by alternate coastdown runs in direction A and B. The resulting resistance w_1 in Newton is calculated using this averaged wind velocity $\overline{v}_{\text{Wind}}$ multiplied with the road load coefficient of second order f_2 (see Equation 2.49). Furthermore, it is assumed that this results in a constant force. Thus, this correction is subtracted only from the constant term f_0 of the road load coefficients. At this point it is clarified that the influence of sidewinds is not considered, although they would also have an impact on the road load in x direction, as explained in section 2.2.1. If the lowest arithmetic average wind velocity is $2\,\text{m/s}$ or less, the wind correction can be left out [10].

- For the correction of the rolling resistance (and the drivetrain losses) the correction term $\left(1 + K_0 \cdot \left(\overline{T}_{\text{Amb}} - T_0\right)\right)$, which is already known from literature [10, 26, 40] (see Equation 2.27), is applied. As the rolling resistance can be described by a polynomial with coefficients of constant, first and fourth order (see Equation 2.23), the correction term is applied to the road load coefficients f_0 and f_1. In the GTR No. 15 a correction factor K_0 with a value of 0.86 % is stated. In the literature [26, 40] a value range between 0.5 % and 0.8 % is stated. However, in the literature only the rolling resistance is corrected, but not additionally the drivetrain losses according to the coastdown method.

- The aerodynamic drag can be described using a polynomial of second order (see Equation 2.11). Therefore, the correction term K_2 is only applied to the road load coefficient of second order f_2. The correction term considers the change of the atmospheric density ρ_{Air} due to different atmospheric pressures and temperatures. The correction is made assuming the ideal gas equation (see Equation 2.12).

Finally, the road load coefficients f_0^c, f_1^c and f_2^c are determined again with a polynomial regression of second order (see Equation 2.2). The road load calculated with these coefficients $F_{j,\text{CDM}}^c$ is defined as:

$$F_{j,\text{CDM}}^c = f_0^c + f_1^c \cdot v_j + f_2^c \cdot v_j^2 \tag{2.50}$$

where:

$F_{j,\text{CDM}}^c$	is the road load of the vehicle at the reference velocity point v_j for the coastdown method calculated on the basis of the corrected road load coefficients f_0^c, f_1^c and f_2^c in N;
f_0^c	is the constant term of the corrected road load coefficients in N;
f_1^c	is the corrected coefficient of the first order term in $\text{N}/(\text{m/s})$;
f_2^c	is the corrected coefficient of the second order term in $\text{N}/(\text{m/s})^2$.

2.3.2 Wind tunnel method according to WLTP

The Wind Tunnel Method (WTM) according to WLTP uses a wind tunnel in combination with a flat belt dynamometer to determine the road load of a vehicle. The method is schematically illustrated in Figure 2.26. Instead of a flat belt dynamometer also a chassis dynamometer can be used. However, only the method in combination with a flat belt dynamometer is explained in detail in the following. At the end of this subsection a short overview of the differences with the method in combination with a chassis dynamometer is given.

According to this method, the aerodynamic drag coefficient c_{w} for the resulting aerodynamic drag $F_{j,\text{Air}}$ is determined in a wind tunnel. The remaining components of the road load (mostly rolling resistance $F_{j,\text{Roll}}$ and drivetrain losses $F_{j,\text{Drive}}$) are measured using a flat belt dynamometer, which measures the front and rear axles simultaneously. The sum of the two separately determined road load components is the road load of a vehicle according to the wind tunnel method $F_{j,\text{WTM}}$ at each reference velocity point v_j [10]. In

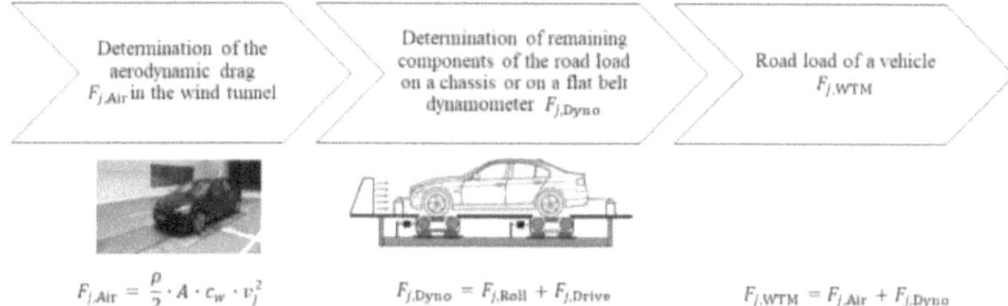

Figure 2.26: Wind tunnel method according to WLTP (adapted by permission from Isabell Vogeler: Different methods for road load determination in comparison: Wind tunnel, Wind tunnel method according to WLTP and Coastdown method [60], 2018).

the following, the test procedure for this measurement method is explained in detail.

The aerodynamic drag $F_{j,\text{Air}}$ of a vehicle is determined with the aerodynamic drag coefficient c_w of a vehicle measured in a wind tunnel according to the following equation [10]:

$$F_{j,\text{Air}} = (c_w \cdot A_x) \cdot \frac{\rho_0}{2} \cdot v_j^2 \tag{2.51}$$

where:

$F_{j,\text{Air}}$	is the calculated aerodynamic drag at the reference velocity point v_j using the aerodynamic drag coefficient c_w measured in the wind tunnel in N;
c_w	is the aerodynamic drag coefficient determined in the wind tunnel;
A_x	is the frontal area of the vehicle in m^2;
ρ_0	is the dry air density and is defined as 1.189 kg/m^3;
v_j	is the reference velocity point j in m/s.

The product of c_w and A_x has to fulfill a precision of $0.015\,m^2$ and the test wind velocity is at least $140\,km/h$ [10].

To determine the remaining components of the road load, the flat belt dynamometer is used. Here, the vehicle is initially accelerated to $80\,km/h$ and then is decelerated with the clutch disengaged or the automatic transmission placed in neutral by moderate braking from $80\,km/h$ to $20\,km/h$ at the test bench. This braking phase is equal to the coastdown method. Afterwards, two different procedures are allowed for the following warm-up phase [10]:

- Standard warm-up (STD): The vehicle drives at 90 % of the maximum velocity of the applicable WLTC. In the case of the WLTC Class 3b, it is $118\,\mathrm{km/h}$.

- Alternative warm-up (ALT): The vehicle is driven by the dynamometer with the clutch disengaged or the automatic transmission placed in neutral at 110 % of the maximum velocity of the applicable WLTC. In the case of the WLTC class 3b, it is $144\,\mathrm{km/h}$.

The warm-up phase lasts for at least $1200\,\mathrm{s}$. If the alternative warm-up is chosen, the duration exceeds $1200\,\mathrm{s}$ until the change of the measured force over a period of $200\,\mathrm{s}$ is less than $5\,\mathrm{N}$. Following the warm-up phase, the measurement phase starts to determine the road load of a vehicle. Here also, two options are allowed [10]:

- Measurement phase with stabilized velocity (SV)

- Measurement phase by deceleration (CD)

These two measurement phases are explained in the following.

Measurement phase with stabilized velocity

During this measurement phase, the vehicle is driven by the dynamometer with the clutch disengaged or the automatic transmission placed in neutral from the highest reference velocity of $130\,\mathrm{km/h}$ to the lowest reference velocity of $20\,\mathrm{km/h}$ in incremental steps of $10\,\mathrm{km/h}$. At each step, the reference velocity is initially stabilized for at least 4 seconds and for a maximum of 10 seconds. Then the force in x direction is measured for at least 6 seconds, while the vehicle velocity is kept constant. The resulting force $F^{*}_{j,\mathrm{Dyno}}$ at each reference velocity v_j is the arithmetic average of the measured forces. Instantly after the measurement time at the current reference velocity point, the dynamometer decelerates from this to the next applicable reference velocity point at $1\,\mathrm{m/s^2}$ [10]. In Figure 2.27 the above described total test procedure with braking phase (wB), standard warm-up phase (STD) and the measurement phase with stabilized velocity (SV) (abbr. SV STD wB) is shown with the velocity profile over time. The same measurement procedure but with the alternative warm-up phase (ALT) (abbr. SV ALT wB) is illustrated in Figure 2.28.

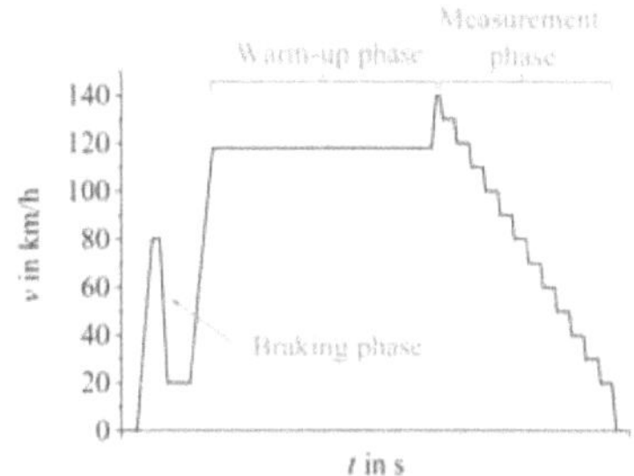

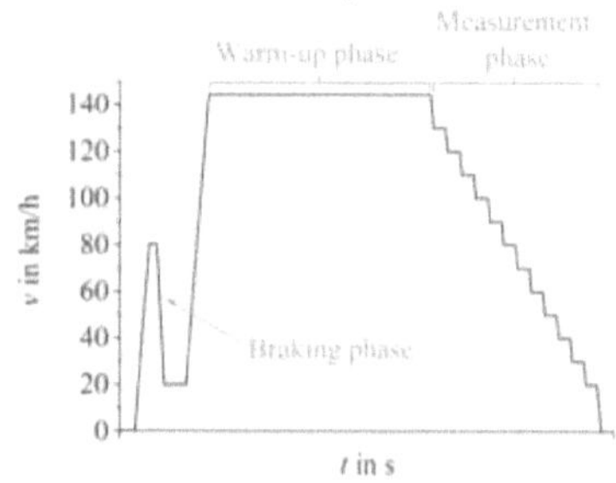

Figure 2.27: Test procedure according to WLTP with braking phase, standard warm-up phase and measurement phase with stabilized velocity (abbr. SV STD wB)

Figure 2.28: Test procedure according to WLTP with braking phase, alternative warm-up phase and measurement phase with stabilized velocity (abbr. SV ALT wB) (adapted by permission from Springer Nature Customer Service Centre GmbH: Springer Nature, 18. Internationales Stuttgarter Symposium [8], 2018).

Measurement procedure by deceleration

If the standard warm-up phase is chosen, the vehicle accelerates after the warm-up phase to 130 km/h and maintains this velocity for at least one minute. If the alternative warm-up phase is used, the vehicle is driven by the dynamometer at 110 % of the highest reference velocity (143 km/h) for at least one minute. Afterwards, the vehicle is accelerated to at least 10 km/h above the highest reference velocity before the measurement phase starts. During the entire measurement phase, the transmission is placed in neutral and any steering movement and vehicle brake system actuation have to be avoided. Equal to the coastdown method on a proving ground, the time, which corresponds to the reference velocity v_j as the elapsed time from the vehicle velocity $(v_j + 1.3\overline{8}\,\text{m/s})$ to $(v_j - 1.3\overline{8}\,\text{m/s})$, is measured. The measurement phase ends after two decelerations, if the force of both decelerations is within $\pm 10\,\text{N}$ at each reference velocity point v_j. Otherwise, at least three coastdowns have to be performed, until the criterion for the statistical precision p_j for each reference velocity point v_j according to Equation 2.32 is fulfilled. The resulting force $F^*_{j,\text{Dyno}}$ at each reference velocity point v_j can be calculated by subtracting the aerodynamic force, which is simulated by the dynamometer during the measurement procedure [10]:

$$F^*_{j,\text{Dyno}} = F_j - c_{\text{d}} \cdot v_j^2 \tag{2.52}$$

where:

$F^*_{j,\text{Dyno}}$	is the uncorrected road load of the vehicle at the reference velocity point v_j determined using the flat belt dynamometer in N;
c_{d}	is the dynamometer set coefficient (see Equation A.6 in the appendix) in $\text{N}/(\text{m/s})^2$;
F_j	is the uncorrected road load of the vehicle at the reference velocity point v_j determined using Equation 2.36 in N.

In Figure 2.29 the above described total test procedure with braking phase (wB), standard warm-up phase (STD) and the measurement phase by deceleration (CD) (abbr. CD STD wB) is shown with the velocity profile over time. The same measurement procedure but with the alternative warm-up phase (ALT) (abbr. CD ALT wB) is illustrated in Figure 2.30.

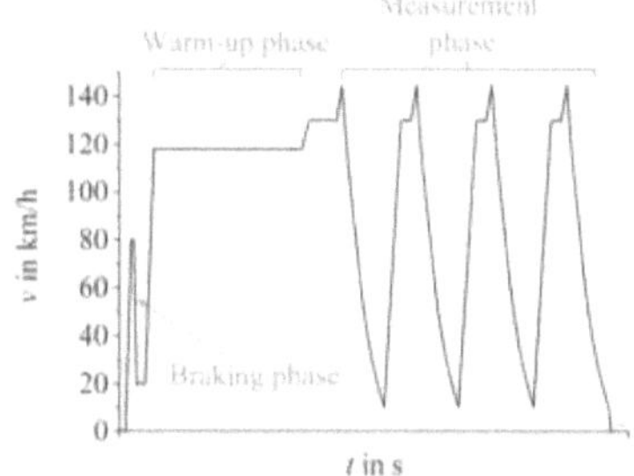

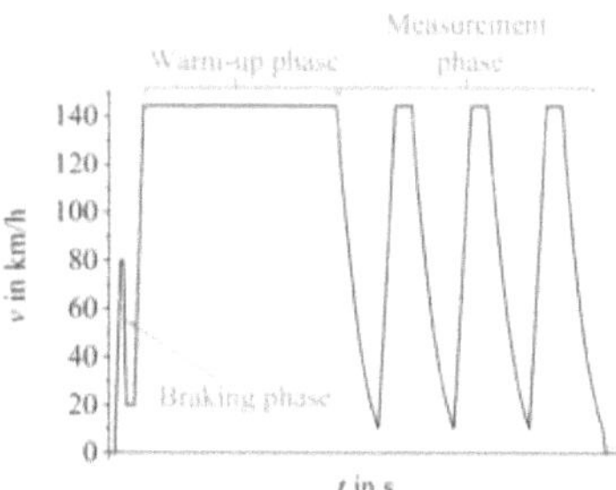

Figure 2.29: Test procedure according to WLTP with braking phase, standard warm-up phase and measurement phase by deceleration (abbr. CD STD wB).

Figure 2.30: Test procedure according to WLTP with braking phase, alternative warm-up phase and measurement phase by deceleration (abbr. CD ALT wB).

Calculation of the corrected road load

Similiar to the coastdown method, the measured force $F^*_{j,\mathrm{Dyno}}$ has to be corrected to reference conditions. However, in this case only the rolling resistance correction factor K_0 and the test mass correction factor K_1 are necessary, as no headwind and tailwind occur on the test benches. The correction equation is defined as [10]:

$$F_{j,\mathrm{Dyno}} = \left(F^*_{j,\mathrm{Dyno}} - K_1 \right) \cdot \left[1 + K_0 \left(\overline{T}_{\mathrm{Dyno}} - 293\,\mathrm{K} \right) \right] \tag{2.53}$$

with

$$K_0 = 8.6 \cdot 10^{-3}\,\mathrm{K}^{-1} \tag{2.54}$$

and

$$K_1 = F^*_{j,\mathrm{Dyno}} \cdot \left(1 - \frac{m_{\mathrm{test}}}{m_{\mathrm{av}}} \right) \tag{2.55}$$

where:

$F_{j,\mathrm{Dyno}}$	is the corrected road load of the vehicle at the reference velocity point v_j measured at the flat belt dynamometer in N;
K_0	is the correction factor for rolling resistance (and drivetrain losses) due to difference in air temperature compared to the reference condition at $20\,^{\circ}\mathrm{C}$ in K^{-1};
K_1	is the test mass correction factor using the differences between the averaged test mass m_{av} compared to the reference test mass m_{Veh} for the chosen vehicle in N;
m_{test}	is the vehicle test mass defined for the road load determination in kg;

m_{av} is the arithmetic average of the test vehicle masses at the beginning and end of the test procedure in kg;

$\overline{T}_{\mathrm{Dyno}}$ is the arithmetic average temperature at the test bench during the measurement procedure in K.

Finally, the road load of a vehicle using the wind tunnel method $F_{j,\mathrm{WTM}}$ is the sum of the two parts $F_{j,\mathrm{Dyno}}$ (see Equation 2.53) and $F_{j,\mathrm{Air}}$ (see Equation 2.51) [10]:

$$F_{j,\mathrm{WTM}} = F_{j,\mathrm{Dyno}} + F_{j,\mathrm{Air}} \tag{2.56}$$

Equal to the coastdown method, the road load coefficients f_0^c, f_1^c and f_2^c for the wind tunnel method are calculated with the least squares regression analysis [10]. The road load $F_{j,\mathrm{WTM}}^c$ can then be calculated on the basis of the road load coefficients equivalent to Equation 2.2.

Wind tunnel method in combination with a chassis dynamometer

The wind tunnel method can also be used in combination with a chassis dynamometer instead of the flat belt dynamometer. The chassis dynamometer is equipped with a single roller at the front and rear axles with a minimum diameter of 1.2 m. However, the forces measured at the chassis dynamomter have to be additionally corrected to a reference, which is equivalent to a road or a flat surface.

For further information about the additional correction terms for the chassis dynamometer see [10].

2.3.3 Further methods for road load determination according to WLTP

In the previous subsections the coastdown method with a stationary anemometer and the wind tunnel method in combination with a flat belt dynamometer were explained in detail. However, there are further methods allowed in the GTR No. 15 [10], which are briefly described in the following. These methods do not only describe the road load determination, but also the determination of the running resistance. The running resistance is defined as the torque resistance measured using torque meters during the forward motion of a vehicle. The torque meters are mounted at the driven wheel of a vehicle [10]. These torque meters are different to the custom-built torque meters introduced in subsection 3.3.4.

Measurement and calculation of the running resistance using the torque meter

Using this method, the running resistance C_j is determined. The torque meters are mounted between the wheel hub and the driven wheel. Before the measurement phase starts, the same braking and warm-up phase are executed as for the coastdown method (see subsection 2.3.1). During the subsequent measurement phase, the wheel torque is measured at the reference velocity points v_j for at least five seconds, while the vehicle keeps the corresponding velocity. The reference velocity points v_j are chosen in descending order from 130 km/h to 20 km/h in incremental steps of 10 km/h. Additionally, during the measurement phase, any steering movement of the wheels and any actuation of the braking system has to be avoided. The measurements are carried out in the opposite direction until for a minimum of three pairs of measurements at each reference velocity v_j a statistical precision of ≤ 0.03 is reached. The statistical precision is calculated similiar to Equation 2.32 for the coastdown method. Instead of the harmonic average of the alternate coastdown time measurements, the arithmetic average of the running resistances C_j for both directions at the reference velocity points v_j are chosen. Finally, the running resistance coefficients are determined using the least square regression method and are corrected to reference conditions, as for the coastdown method. However, due to the installed torque meters, a further correction is necessary, which takes into account the measurement equipment mounted outside on the vehicle, and in particular, its aerodynamic characteristics. For this correction the product of the aerodynamic coefficient multiplied by the frontal area of the vehicle with and without the mounted torque measurement equipment is necessary. Furthermore, a compensation of the drivetrain losses is needed, as they are not contained in these measurement results [10].

Method for the calculation of road load or running resistance based on vehicle parameters

According to this method, the road load or running resistance of a vehicle is intially determined according to the previously described coastdown or torque meter method. If the vehicle is a representative vehicle of a road load matrix family, the road load for an individual vehicle in this family can be calculated based on vehicle parameters such as the test mass m_{test}, the frontal area A_x and the tyre rolling resistance R_{Tyre}. Vehicles are in the same road load matrix family, if they are identical with respect to the following characteristics [10]:

- Transmission type (for example manual, automatic or continuously variable transmission, ...)

- Number of powered axles

Furthermore, a vehicle which fulfills the following criterion is defined as the representative vehicle of the road load matrix family [10]:

- Representative in terms of the estimated poorest aerodynamic drag coefficient and the body shape

- Representative of the estimated average mass of optional equipment

The equation for the calculated road load force $F_{j,c}$ is given as [10]:

$$F_{j,c} = f_0 + f_1 \cdot v_j + f_2 \cdot v_j^2 \tag{2.57}$$

with

$$f_0 = \text{Max}\left\{ 0.05 \cdot f_{0,r} + 0.95 \cdot \left(f_{0,r} \cdot \frac{m_{test}}{m_{test,r}} + \frac{R_{Tyre} - R_{Tyre,r}}{1000} \cdot g_{Earth} \cdot m_{test} \right), \\ 0.2 \cdot f_{0,r} + 0.8 \cdot \left(f_{0,r} \cdot \frac{m_{test}}{m_{test,r}} + \frac{R_{Tyre} - R_{Tyre,r}}{1000} \cdot g_{Earth} \cdot m_{test} \right) \right\} \tag{2.58}$$

$$f_1 = 0 \tag{2.59}$$

and

$$f_2 = \text{Max}\left\{ 0.05 \cdot f_{2,r} + 0.95 \cdot f_{2,r} \cdot \frac{A_x}{A_{x,r}}, \\ 0.2 \cdot f_{2,r} + 0.8 \cdot f_{2,r} \cdot \frac{A_x}{A_{x,r}} \right\} \tag{2.60}$$

where:

$F_{j,c}$	is the calculated road load force for an individual vehicle in a road load matrix family at the reference velocity point v_j in N;
f_0	is the constant road load coefficient for an individual vehicle in a road load matrix family determined using Equation 2.58 in N;
$f_{0,r}$	is the constant road load coefficient for the representative vehicle in a road load matrix family in N;
m_{test}	is the test mass of an individual vehicle in the road load matrix family in kg;
$m_{test,r}$	is the test mass of the representative vehicle in the road load matrix family in kg;
R_{Tyre}	is the tyre rolling resistance of an individual vehicle in the road load matrix family in kg/t;
$R_{Tyre,r}$	is the tyre rolling resistance of the representative vehicle in the road load matrix family in kg/t;
g_{Earth}	is the earth's gravity and set to 9.81 m/s²;
f_1	is the coefficient of the first order term of the road load coefficients in N/(m/s) and is set to zero in this case (see Equation 2.59);
v_j	is the reference velocity point in m/s;
f_2	is the coefficient of the second order term of the road load coefficients for an individual vehicle in the road load matrix family determined using Equation 2.60 in N/(m/s)²;

$f_{2,r}$ is the coefficient of the second order term of the road load coefficients for the representative vehicle in the road load matrix family in $^N/_{(m/s)^2}$;

A_x is the frontal area of an individual vehicle in the road load matrix family in m^2;

$A_{x,r}$ is the frontal area of the representative vehicle in the road load matrix family in m^2.

Instead of the road load coefficients f_0, f_1 and f_2, also the running resistance coefficients c_0, c_1 and c_2, determined with the torque meter method for the representative vehicle in a road load matrix family (see subsection 2.3.3), can be used to calculate the running resistance for an individual vehicle. In this case, besides the dynamic rolling radius determined at $80\,^{km}/_h$, also a coefficient with a value of 1.02 is necessary to compensate for the drivetrain losses of a vehicle. The equations for this calculation can be taken from [10].

Calculation of the road load based on vehicle parameters

A further method is to calculate the road load of a vehicle based on vehicle parameters such as the test mass m_{test}, the vehicle height h_{Veh} and the vehicle width w_{Veh} with different estimation factors. The resulting road load $F_{j,cd}$ can be calculated using the following equations [10]:

$$F_{j,cd} = f_0 + f_1 \cdot v_j + f_2 \cdot v_j^2 \tag{2.61}$$

with

$$f_0 = 0.140\,^{m}/_{s^2} \cdot m_{test} \tag{2.62}$$

$$f_1 = 0 \tag{2.63}$$

and

$$f_2 = \left(3.6 \cdot 10^{-5}\,^{m/s^2}/_{(m/s)^2} \cdot m_{test}\right) + \left(0.2203\,^{kg/m\,s^2}/_{(m/s)^2} \cdot h_{Veh} \cdot w_{Veh}\right) \tag{2.64}$$

where:

$F_{j,cd}$ is the calculated road load force based on the vehicle parameters test mass, vehicle height and width at the reference velocity point v_j in N;

h_{Veh} is the vehicle height as defined in Standard ISO 612:1978 in m;

w_{Veh} is the vehicle width as defined in Standard ISO 612:1978 in m.

2.4 Cycle energy demand, difference in cycle energy demand and driving cycles

To compare the previously described methods for road load determination, the differences in cycle energy demand, for example, between the wind tunnel method and and the coastdown method are evaluated. In the first part of this section, the meaning and the calculation of the cycle energy demand and of the difference in cycle energy demand is explained. Subsequently, the driving cycles, which are used to compare the methods for road load determination, are introduced.

2.4.1 Calculation of cycle energy and cycle energy difference

The cycle energy term is understood to mean the positive energy required by the vehicle to drive a prescribed test cycle. The test cycles used for this comparison are introduced in the next subsection 2.4.2.

In general, the total energy demand can be described as a work W which has to be applied by the vehicle to drive the driving cycle. The work is defined as the the following integral:

$$W = \int F \, dd = E \tag{2.65}$$

To simplify the calculation of the total energy demand E, the energy demand during a specific time period E_i over the corresponding cycle time between t_{start} and t_{end} is summed [10]:

$$E = \sum_{t_{\text{start}}}^{t_{\text{end}}} E_i \tag{2.66}$$

where:

W	is the work applied by the vehicle to drive in J;
F	is a force in N;
d	is a distance in m;
E	is the total energy demand for a vehicle in J;
t_{start}	is the time, at which the chosen test cycle starts, in s;
t_{end}	is the time, at which the chosen test cycle ends, in s;
E_i	is the energy demand of a vehicle during a specific time period $(i-1)$ to i in J.

whereby E_i is defined as follows, if the vehicle is accelerating or maintains a constant velocity [10]:

$$E_i = F_i \cdot d_i \text{ if } F_i > 0 \tag{2.67}$$

But if the vehicle is decelerated or stationary, E_i is defined as [10]:

$$E_i = 0 \text{ if } F_i \leq 0 \tag{2.68}$$

where:

F_i is the force for driving the vehicle during the time period $(i-1)$ to i in N;

d_i is the distance the vehicle is travelling during the time period $(i-1)$ to i in m.

The force F_i which is needed to move the vehicle during the time period $(i-1)$ to i can be calculated with the following equation [10]:

$$F_i = f_0^c + f_1^c \cdot \left(\frac{v_i + v_{i-1}}{2} \right) + f_2^c \cdot \frac{(v_i + v_{i-1})^2}{4} + (k_r \cdot m_\text{test}) \cdot a_i \qquad (2.69)$$

where:

v_i is the velocity at the time step t_i in $\mathrm{m/s}$;

k_r is the factor which considers the inertial resistances of the drivetrain during acceleration and deceleration and is set to 1.03;

m_test is the vehicle test mass defined for the road load determination in kg;

a_i is the acceleration of the vehicle during the time period $(i-1)$ to i in $\mathrm{m/s^2}$;

f_0^c, f_1^c, f_2^c are the corrected road load coefficients for the tested vehicle according to the coastdown method (see section 3.1.1), wind tunnel method (see section 3.1.2) or AEROLAB method (see section 3.1.3) in N, in $\mathrm{N/(m/s)}$ and in $\mathrm{N/(m/s)^2}$, respectively.

The acceleration of the vehicle a_i is defined as [10]:

$$a_i = \frac{v_i - v_{i-1}}{t_i - t_{i-1}} \qquad (2.70)$$

where:

t_i is the time step in s.

The time increment for the investigated driving cycles (see subsection 2.4.2) is 1 second. To show that the simplification made in Equation 2.66, to avoid a calculation with an integral (Equation 2.65), is valid, the time increment of 1 second is varied and the velocity profile is then interpolated by a spline. Afterwards, the resulting change in cycle energy demand related to the cycle energy demand with a time increment of 1 second is calculated. In Figure 2.31 it can be seen that a further decrease of the time increment results in a maximum difference of about 0.2 % of the cycle energy demand as compared to the cycle energy demand with a time increment of 1 second. In contrast, in case that the time step t_i is increased to 2 seconds the resulting difference in energy demand is about 0.7 %. This clarifies that the simplification made in Equation 2.66 is valid and the simplification can be applied.

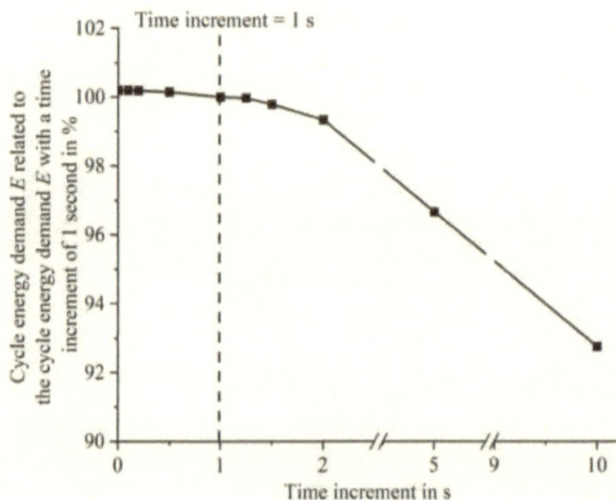

Figure 2.31: Change of the cycle energy demand depending on the time increment related to the cycle energy demand with a time increment of 1 second.

For the further comparison of the different methods for road load determination, the difference in cycle energy demand ϵ, for example, between the wind tunnel method (WTM) and the coastdown method (CDM) is used. It is defined as follows [10]:

$$\epsilon_{\mathrm{WTM}} = \frac{E_{\mathrm{WTM}}}{E_{\mathrm{CDM}}} - 1 \tag{2.71}$$

where:

ϵ_{WTM}	is the difference in cycle energy demand;
E_{WTM}	is the total cycle energy demand over a complete test cycle with the road load determined using the wind tunnel method (WTM) in J;
E_{CDM}	is the total cycle energy demand over a complete test cycle with the road load determined using the coastdown method (CDM) in J.

In the following, the difference in cycle energy demand ϵ is always given in percent.

2.4.2 Driving cycles

In this study, different methods for the road load determination are compared using the ARTEMIS (Assessment and Reliability of Transport Emission Models and Inventory Systems) European driving cycle[9] and the WLTC (Class 3b) [10]. These two driving cycles are illustrated in Figure 2.32, where the velocity profiles are plotted over the cycle times.

---

[9]The source of the database for the velocity-time-profile of the ARTEMIS drving cycle is the Department of Type Approval and Emissions of the BMW Group.

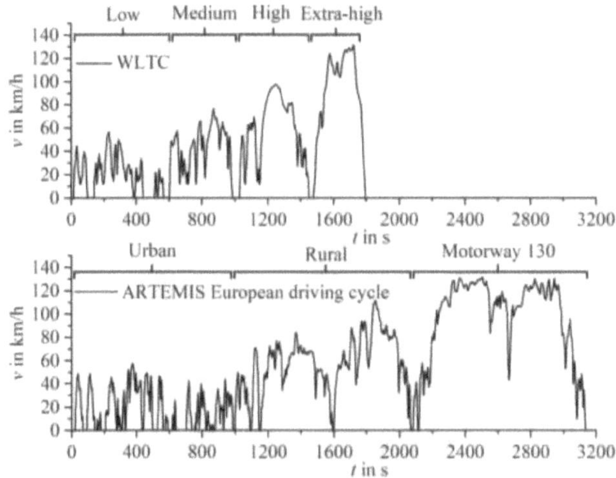

Figure 2.32: Velocity over time profile for WLTC [10] and ARTEMIS driving cycle[9].

ARTEMIS European driving cycle

In the framework of the ARTEMIS European research project, a set of reference real-world driving cycles have been developed. Therefore, previously recorded, real driving data have been used as a database for the ARTEMIS cycles. The first data set is taken from MODEM (MODelling of EMissions and fuel consumption in urban areas), which is a research project within the DRIVE initiative (Dedicated Road Infrastructure for Vehicle safety in Europe). For this research project 58 representative European private vehicles equipped with an on-board data collection system were driven for their normal purpose by their owners in France, the UK and Germany. Then, this data set is enlarged by the HYZEM (European development of HYbrid vehicle technology approaching efficient Zero EMission mobility) research project, where further 19 private vehicles were monitored in Greece, using the same methodology as for MODEM. In total, the database for the ARTEMIS cycles consists of 2,000 days, which is equal to 5.5 years, 10,300 trips, 88,000 km travelling and 2,200 hours of driving. Subsequently, the data set was validated through two further measurement programs in Italy and Switzerland [62].

Furthermore, the ARTEMIS European driving cycle is a 'multi-component' driving cycle, which includes an urban, a rural road and a motorway section, which allows a separation of the emissions according to more specific driving conditions. The different situations are described in sub-cycles, which can be combined into one single cycle, as shown in Figure 2.32 [62].

Additionally, in Table 2.5 characteristics such as cycle duration, distance travelled, average velocity, maximum velocity, velocity distribution and average and maximum acceleration for the single sections of the ARTEMIS European driving cycle are listed, whereby the

Table 2.5: Characteristics of the ARTEMIS European driving cycle sections (urban, rural road, motorway 130)

Characteristic	Urban	Rural road	Motorway 130
Cycle duration in s	993	1082	1068
Distance travelled in m	4870	17272	28736
Average velocity in $\mathrm{km/h}$	17.7	57.5	96.9
Maximum velocity in $\mathrm{km/h}$	57.7	111.5	131.8
Velocity distribution in %			
- Idle ($v = 0\,\mathrm{km/h}$)	28.4	3.0	1.5
- Low velocity ($0\,\mathrm{km/h} < v \leq 50\,\mathrm{km/h}$)	69.3	31.3	15.1
- Medium velocity ($50\,\mathrm{km/h} < v \leq 90\,\mathrm{km/h}$)	2.3	58.5	13.3
- High velocity ($v > 90\,\mathrm{km/h}$)	0.0	7.1	70.1
Average acceleration in $\mathrm{m/s^2}$	0.7	0.5	0.4
Maximum acceleration in $\mathrm{m/s^2}$	2.9	2.4	1.9

average acceleration is determined without zero values[9].

It is shown that the sections rural road and motorway 130 with velocities over $100\,\mathrm{km/h}$ are the longest parts in the ARTEMIS European driving cycle. In the urban section 69.3 % of the velocities are in the low velocity range ($0\,\mathrm{km/h} < v \leq 50\,\mathrm{km/h}$). In addition, there is a large part of 28.4 % where the vehicle is just idling. However, both the average and the maximum acceleration with $0.7\,\mathrm{m/s^2}$ and $2.9\,\mathrm{m/s^2}$ in this section are the highest values compared to the other sections. The rural road section consists mainly of driving in the low velocity range ($0\,\mathrm{km/h} < v \leq 50\,\mathrm{km/h}$) with 31.3 % and the medium velocity range ($50\,\mathrm{km/h} < v \leq 90\,\mathrm{km/h}$) with 58.5 %. In contrast, in the Motorway 130 section the high velocity range ($v > 90\,\mathrm{km/h}$) is the largest part with 70.1 %. The average velocities are in stages from $17.7\,\mathrm{km/h}$ to $96.9\,\mathrm{km/h}$ over the sections. And a velocity over $100\,\mathrm{km/h}$ is already reached in the rural road section [63].

WLTC

The driving cycle WLTC is defined as a new test procedure, which was introduced in the EU and in various other countries in 2017. The development was carried out by the working party on pollution and energy transport of the World Forum for the Harmonization of Vehicle Regulation of the UN-ECE (United Nations Economic Commission for Europe). The approach of WLTC is to represent typical driving characteristics around the world. For this driving behaviour, data based on statistical information about light vehicles used in different regions of the world was collected and analyzed. The chosen regions are Germany, Spain, Italy, Poland, Slovenia, United Kingdom, Belgium, France, Sweden, Switzerland, USA, Japan, Korea and India. For the driving behaviour over 765,000 km of data of different vehicle categories (like various engine capacities, power-to-mass ratios, manufacturers etc.) on different road types (urban, rural, motorway) and driving conditions (peak, off-peak and weekend) is collected. In contrast to the ARTEMIS driving cycle, the WLTC is not classified according to road categories like urban, rural road and motorway, as there are too many differences in definitions and velocity limits for different regions, as can be seen in Table 2.6 [64].

It can be seen that in India, Korea and Japan, the maximum velocity limit is in the range from $40\,\mathrm{km/h}$ to $80\,\mathrm{km/h}$, depending on the kind of urban area. On rural roads, the maximum velocity limit is $80\,\mathrm{km/h}$ in Korea, whereby in India and Japan the maximum is $60\,\mathrm{km/h}$. However, the highest differences are found for the road category motorway. Here, the maximum allowed velocity is $80\,\mathrm{km/h}$ in India, $120\,\mathrm{km/h}$ in Korea and $100\,\mathrm{km/h}$ in Japan. Additionally, even in the European Union (EU), for all three road categories, there are different velocity limits depending on the regulations in the individual countries. Therefore, a driving cycle was developed, which is based on velocity classes (low, medium, high velocity) instead of road categories (urban, rural, motorway). In addition, the high velocity phase is split into two further sections: a high and an extra-high velocity section, whereby the high velocity section represents the driving behaviour in Asian regions and the extra-high velocity section is more characteristic of European and USA driving behaviour [64].

The test cycle duration is set to $1800\,\mathrm{s}$, which is both statistically representative and feasible to test in the laboratory. The length of each velocity phase is defined, based on the traffic volume between the low, medium, high and extra-high phases [64].

Table 2.6: Definition of road categories and velocity limits in different regions [64]

Region	Urban	Rural road	Motorway
India	Paved roads with velocity limit of $\leq 40\,\mathrm{km/h}$; (mountain areas not included)	Paved non-motorways with a velocity limit between $40\,\mathrm{km/h}$ and $60\,\mathrm{km/h}$ in urban and non-urban areas	Paved motorways and multi-lane roads for fast traffic with a velocity limit range from $60\,\mathrm{km/h}$ to $80\,\mathrm{km/h}$
Korea	Arterial, collector and local road in and near business district areas with a velocity limit range from $40\,\mathrm{km/h}$ to $80\,\mathrm{km/h}$ depending on the road type	Arterial, collector and local road in non-urban areas with velocity limit range from $50\,\mathrm{km/h}$ to $80\,\mathrm{km/h}$ depending on the road type	Motorway for fast traffic in both urban and non-urban areas with a velocity limit range from $100\,\mathrm{km/h}$ to $120\,\mathrm{km/h}$ depending on the area
Japan	Velocity limit of $\leq 60\,\mathrm{km/h}$ in densely populated areas; mountain areas not included	Non-motorways and non-densely populated areas with a velocity limit of $\leq 60\,\mathrm{km/h}$; mountain areas not included	Motorways in urban and between urban areas with a velocity limit $\leq 100\,\mathrm{km/h}$; mountain areas not included
EU	Depending on the resprective legal framework of the different EU member states	Depending on the resprective legal framework of the different EU member states	Depending on the resprective legal framework of the different EU member states

In Table 2.7 the WLTC Class 3b is analyzed due to the same characteristics as used for the ARTEMIS European driving cycle (see Table 2.5). Class 3b means that the vehicles to be tested have a maximum velocity higher than $120\,\mathrm{km/h}$.

Table 2.7: Characteristics of the WLTC Class 3b sections (low, medium, high and extra-high velocity phases)

Characteristic	Low	Medium	High	Extra-High
Cycle duration in s	589	433	455	323
Distance travelled in m	3095	4756	7162	8254
Average velocity in $\mathrm{km/h}$	18.9	39.4	56.5	92.0
Maximum velocity in $\mathrm{km/h}$	56.5	76.6	97.4	131.3
Velocity distribution in %				
- Idle ($v = 0\,\mathrm{km/h}$)	25.3	11.3	6.8	2.2
- Low velocity ($0\,\mathrm{km/h} < v \leq 50\,\mathrm{km/h}$)	71.8	51.5	30.5	11.8
- Medium velocity ($50\,\mathrm{km/h} < v \leq 90\,\mathrm{km/h}$)	2.9	37.4	47.5	23.2
- High velocity ($v > 90\,\mathrm{km/h}$)	0.0	0.0	15.4	62.8
Average acceleration in $\mathrm{m/s^2}$	0.5	0.4	0.4	0.3
Maximum acceleration in $\mathrm{m/s^2}$	1.6	1.6	1.7	1.1

It can be seen that in sum the high and extra-high section is the longest part of the WLTC, referenced to the distance travelled. Similar to the ARTEMIS European driving cycle, the parts with a velocity of 0 $\mathrm{km/h}$ can be found mostly in the low velocity sections (low and medium section for WLTC and urban section for ARTEMIS European driving cycle). The average velocities are in stages from $18.9\,\mathrm{km/h}$ to $92.0\,\mathrm{km/h}$ over the sections. But a velocity over $100\,\mathrm{km/h}$ is reached first in the extra-high section. In contrast, in the ARTEMIS European driving cycle a velocity of 111.5 $\mathrm{km/h}$ is already reached in the second section (rural road). Furthermore, the acceleration values are at their highest in the low velocity section on average, similar to the ARTEMIS European driving cycle.

However, in this study the different methods for road load determination are compared due to the cycle energy demand of the entire cycle instead of those of the individual sections. Therefore, the following subsection compares the whole cycles to each other.

Comparison of ARTEMIS driving cycle and WLTC

As already mentioned, in Figure 2.32 the velocity profiles of the WLTC (top) and of the ARTEMIS European driving cycle (bottom) are plotted over time. Additionally, in Table 2.8 the individual cycle characteristics for the entire drivig cycles are listed. It can be seen that the ARTEMIS European driving cycle is about 47 % longer than WLTC. On the other hand, the maximum velocities of 131.8 $^{\mathrm{km}}/_{\mathrm{h}}$ and 131.3 $^{\mathrm{km}}/_{\mathrm{h}}$ are nearly the same. However, due to the higher average velocity of 58.3 $^{\mathrm{km}}/_{\mathrm{h}}$ (ARTEMIS) as compared to 46.5 $^{\mathrm{km}}/_{\mathrm{h}}$ (WLTC) the distance travelled using the AERTEMIS European driving cycles is about 54 % longer than the one for WLTC. Furthermore, the part of idle ($v = 0\,^{\mathrm{km}}/_{\mathrm{h}}$) is 2.5 % lower for ARTEMIS than for WLTC. Instead, the high velocity part ($v > 90\,^{\mathrm{km}}/_{\mathrm{h}}$), reported as being 26.3 %, is significantly greater when compared to WLTC with a value of 15.2 %. Moreover, in the ARTEMIS European driving cycle there is also a higher averaged and significantly higher maximum acceleration when compared to WLTC.

Table 2.8: Characteristics of ARTEMIS European driving cycle and WLTC

Characteristic	WLTC (Class 3b)	ARTEMIS
Cycle duration in s	1800	3143
Distance travelled in m	23266	50878
Average velocity in $^{\mathrm{km}}/_{\mathrm{h}}$	46.5	58.3
Maximum velocity in $^{\mathrm{km}}/_{\mathrm{h}}$	131.3	131.8
Velocity distribution in %		
- Idle ($v = 0\,^{\mathrm{km}}/_{\mathrm{h}}$)	13.0	10.5
- Low velocity ($0\,^{\mathrm{km}}/_{\mathrm{h}} < v \leq 50\,^{\mathrm{km}}/_{\mathrm{h}}$)	45.7	37.8
- Medium velocity ($50\,^{\mathrm{km}}/_{\mathrm{h}} < v \leq 90\,^{\mathrm{km}}/_{\mathrm{h}}$)	26.1	25.4
- High velocity ($v > 90\,^{\mathrm{km}}/_{\mathrm{h}}$)	15.2	26.3
Average acceleration in $^{\mathrm{m}}/_{\mathrm{s}^2}$	0.4	0.5
Maximum acceleration in $^{\mathrm{m}}/_{\mathrm{s}^2}$	1.7	2.9

2.5 Exemplary calculation of the road load

To gain an impression of the composition of the road load, the road load of a fictional vehicle is calculated using values from the literature. In the following, a higher mid-class vehicle with the characteristics listed in Table 2.9 is assumed.

The aerodynamic drag $F_{j,\mathrm{Air}}$ is calculated using Equation 2.11 with an air density value of 1.189 $^{\mathrm{kg}}/_{\mathrm{m}^3}$. The resulting values for the velocity range from 130 $^{\mathrm{km}}/_{\mathrm{h}}$ to 20 $^{\mathrm{km}}/_{\mathrm{h}}$ are given in Table 2.10, respectively.

The rolling resistance F_{Roll} is estimated using Equation 2.15 with a rolling resistance coefficient of 0.01 for fine asphalt. Furthermore, it is assumed that the rolling resistance with the resulting value of 196.2 N is constant over the velocity range from 20 $^{\mathrm{km}}/_{\mathrm{h}}$ to 130 $^{\mathrm{km}}/_{\mathrm{h}}$.

Table 2.9: Exemplary calculation of the road load of a fictional vehicle

Characteristic	Value	Source
Vehicle class	higher mid-class	-
Wheel drive	rear wheel drive (RWD)	-
Vehicle mass m_{Veh} in kg	2000	-
Frontal area A_x in m^2	2.3	[20]
Aerodynamic coefficient c_x	0.253	[20]
Rolling resistance coefficient f_{R} for fine asphalt	0.010	[16]
Wheel ventilation coefficient c_{Vent} for a rim with 5 spokes	0.0074	[29]
Lift force coefficient c_{A}	0.125	[20]
Camber angle γ in degrees	-2	[17]
Camber stiffness in N/°	220	see Figure 2.16
Total toe angle δ_0 at the front axle for a vehicle with RWD in '	+20	[17]
Total toe angle δ_0 at the rear axle in '	+20	[17]
Cornering stiffness in N/°	2410	see Figure 2.16

However, it should be considered that the relationship between the rolling resistance coefficient and the velocity can be approximated as a 4th order polynomial (see in Figure 2.15, Equation 2.23 and [22]). In addition, it is clarified that due to the value range of the rolling resistance coefficient on fine asphalt of 0.009 to 0.011, the rolling resistance for the chosen test weight of 2000 kg is differing from 176.6 N to 215.8 N. Furthermore, due to lift and down forces acting on the vehicle the wheel load F_{N} is reduced or increased, respectively. For this case, a lift force coefficient c_{A} of 0.125 is assumed at both the front and the rear axle (compare the average value for a higher midclass vehicle in [20]). The resulting lift forces reducing the resulting wheel load F_{N} and in consequence the resulting rolling resistance $F_{j,\mathrm{Roll,Lift}}$ can be calculated using the following equations [20]:

$$F_{j,\mathrm{Roll,Lift}} = (F_{\mathrm{N}} - F_{j,\mathrm{A}}) \cdot f_{\mathrm{r}} \tag{2.72}$$

with

$$F_{j,\mathrm{A}} = c_{\mathrm{A}} \cdot A_x \cdot \frac{\rho_{\mathrm{Air}}}{2} \cdot v_j^2 \tag{2.73}$$

where:

$F_{j,\mathrm{Roll,Lift}}$	is the rolling resistance reduced by lift forces in N;
F_{N}	is the wheel load of the vehicle in N;
f_{r}	is the rolling resistance coefficient;
$F_{j,\mathrm{A}}$	is the lift/downward force in N;
c_{A}	is the lift/downward force coefficient;
A_x	is the frontal area of the vehicle in m^2;

ρ_{Air} is the air density in $\mathrm{kg/m^3}$;

v_j is the reference velocity point in $\mathrm{m/s}$.

The resulting reduced rolling resistance is stated in Table 2.10. In the next step, a total toe angle δ_0 at the front and at the rear angle of 20' and a camber angle γ of -2° for all tyres are assumed. The additional resistance $F_{\text{Roll},\alpha}$ is determined with Equation 2.22. The wheel ventilation resistance $F_{j,\text{Vent}}$ is approximated using the results given in Link [29]. For the fictional vehicle wheel rims with 5 spokes (see Setup 1 in [29]) with a wheel ventilation coefficient c_{Vent} of 0.0074 are assumed. The additional wheel ventilation resistance $F_{j,\text{Vent}}$ is calculated using the same equation as the one for the aerodynamic drag (Equation 2.11). However, instead of the aerodynamic drag coefficient c_{W}, the wheel ventilation coefficient c_{Vent} is used in this case.

Finally, the drivetrain losses are estimated with 50 N, which also includes residual brake forces [26]. Therefore, the calculated road load $F_{j,\text{Calc}}$ is determined as follows:

$$F_{j,\text{Calc}} = F_{j,\text{Air}} + F_{j,\text{Roll,Lift}} + F_{\text{Roll},\alpha} + F_{j,\text{Vent}} + F_{j,\text{Drive}} \qquad (2.74)$$

where:

$F_{j,\text{Calc}}$ is the calculated road load of a fictional vehicle in N;

$F_{\text{Roll},\alpha}$ is the resistance due to tyre slip angle α in N;

$F_{j,\text{Vent}}$ is the wheel ventilation resistance in N;

$F_{j,\text{Drive}}$ are the drivetrain losses in N.

In Figure 2.33 the calculated road load $F_{j,\text{Calc}}$ (black line with square) and its components (stated in Table 2.10) are plotted over the reference velocity points v_j. Additionally, the ratios of the single components related to the calculated road load for all reference velocity points v_j are illustrated in Figure 2.34.

Table 2.10: Exemplary calculation of the road load $F_{j,\text{Calc}}$ of a fictional vehicle

v_j in km/h	$F_{j,\textbf{Air}}$ in N	$F_{j,\textbf{Roll,Lift}}$ in N	$F_{\textbf{Roll},\alpha}$ in N	$F_{j,\textbf{Vent}}$ in N	$F_{j,\textbf{Drive}}$ in N	$F_{j,\textbf{Calc}}$ in N
20	10.6	196.1	9.8	0.3	50	266.9
30	23.9	196.1	9.8	0.7	50	280.5
40	42.5	196.0	9.8	1.2	50	299.5
50	66.4	195.9	9.8	1.9	50	324.0
60	95.6	195.7	9.8	2.8	50	353.9
70	130.1	195.6	9.8	3.8	50	389.3
80	170.0	195.4	9.8	5.0	50	430.1
90	215.1	195.1	9.8	6.3	50	476.3
100	265.6	194.9	9.8	7.8	50	528.0
110	321.4	194.6	9.8	9.4	50	585.2
120	382.4	194.3	9.8	11.2	50	647.7
130	448.8	194.0	9.8	13.1	50	715.7

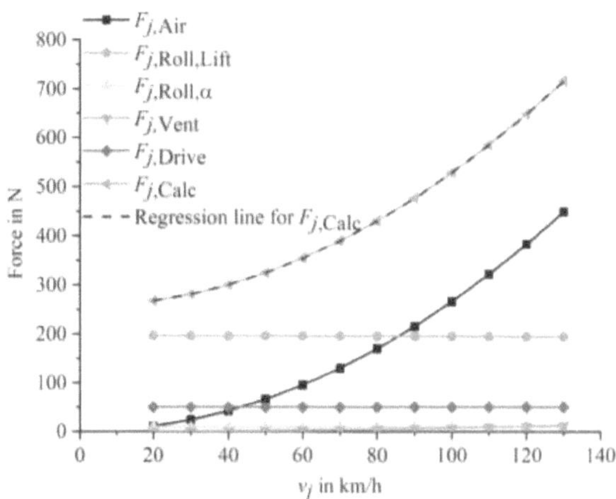

Figure 2.33: Calculated road load $F_{j,\text{Calc}}$ and its components of the reference velocity points v_j.

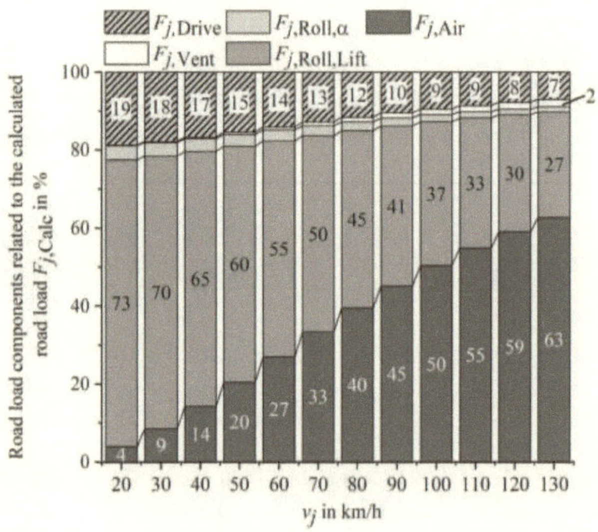

Figure 2.34: Ratios of the road load components referenced to the total calculated road load $F_{j,\text{Calc}}$ for the reference velocity points v_j.

It can be seen that until a velocity of $70\,\text{km/h}$ the rolling resistance $F_{j,\text{Roll,Lift}}$ is the major part of the total road load. And only from a velocity of $100\,\text{km/h}$ the aerodynamic drag is dominating. The ratio of the drivetrain losses $F_{j,\text{Drive}}$ decreases from $19\,\%$ to $7\,\%$ with increasing velocity. The highest ratio of $2\,\%$ for the wheel ventilation resistance $F_{j,\text{Vent}}$ occurs at $130\,\text{km/h}$. The ratio for the resistance due to toe and camber $F_{\text{Roll},\alpha,\gamma}$ is even lower.

As already stated in Equation 2.2 (see chapter 2.1) the road load curve can be described as a polynomial of second order, which is also illustrated as the regression line (black dashed line) in Figure 2.34:

$$F_{j,\text{Calc}} = f_0 + f_1 \cdot v_j + f_2 \cdot v_j^2 \tag{2.75}$$

where:

$f_0,\ f_1,\ f_2$ are the road load coefficients describing the road load curve as a polynomial of second order in N, in N/(m/s) and in N/(m/s)^2, respectively.

The coefficients can be determined using the least square method [14]. As it can be seen in Figure 2.33 and from the related equations (see Equation 2.11), the aerodynamic drag $F_{j,\text{Air}}$, the wheel ventilation resistance $F_{j,\text{Vent}}$ and also the lift forces (see Equations 2.73 and 2.72) depend quadratically on the velocity v_j. Therefore, these road load components mainly define the road load coefficient f_2. In contrast, the rolling resistance is mainly described by the road load coefficients f_0 and f_1 as can also be seen in Equation 2.23. In the case that the drivetrain losses are constant, as stated in this example, this resistance is described by the coefficient f_0. However, if the drivetrain losses also depend on the velocity (see for example in Figure 4.15), these losses are also described by the road load coefficient f_1. The resulting road load coefficients, describing the calculated road load $F_{j,\text{Calc}}$, are given in Table 2.11.

Table 2.11: Road load coefficients f_0, f_1 and f_2 for the calculated road load $F_{j,\text{Calc}}$

f_0 in N	f_1 in $\mathrm{N}/\mathrm{(m/s)}$	f_2 in $\mathrm{N}/\mathrm{(m/s)^2}$
256.0	0.0	0.3525

With these road load coefficients the total energy demands E for both driving cycles WLTC and the ARTEMIS are calculated using Equation 2.66. The results are given in Table 2.12. Comparing the results, it is pointed out that the theoretical calculated energy demand per kilometer is 12.4 % higher for ARTEMIS driving cycle. One reason is the higher proportion of velocities, which are greater than $90\,\mathrm{km/h}$. The proportion for this velocity range is 26.3 % for the ARTEMIS cycle and only 15.2 % for WLTC. A further reason for this are the higher averaged and absolute acceleration values for the ARTEMIS driving cycle (see Table 2.8 in subsection 2.4.2).

Table 2.12: Total energy demand E and total energy demand per kilometer E_{km} for the driving cycles WLTC and ARTEMIS of the fictional vehicle

	WLTC	ARTEMIS
E in kJ	14,527.3	35,700.0
E per kilometer in $\mathrm{kJ/km}$	0.624	0.702

2.6 Error calculation according to GUM

The uncertainty of a measurement is characterized by the quality of the measurement results. Its estimation is essential for the comparability and the acceptance of the measurement results, and thus, for decisions to be taken based on these. Hence, calibration and measurement results are only complete and acceptable in combination with an uncertainty statement [65, 66]. The Guide to the Expression of Uncertainty in Measurement (GUM) is a worldwide standard for the evaluation and expression of measurement uncertainty. It is based on the knowledge of the measurement procedure, the evaluation of the input quantities and the modelling of the measurement [67]. In the following, the concept of GUM is explained in more detail.

The value y of the measurement object Y can be determined by a measurement, whereby the value is the particular quantity of the object. However, the measurand Y is usually not measured directly, instead it is determined from N other input quantities X_1, X_2, ..., X_N. Examples for further input quantities are ambient atmospheric temperature or device parameters. The relationship between the measurand Y and the input quantities X_N can be described by a functional relationship f [67, 68]:

$$Y = f(X_1, X_2, ..., X_N) \text{ with } N = 1, 2, 3, ... \tag{2.76}$$

where:

Y	is the measurand;
f	is the functional relationship of the input quantities X_m;
X_N	is an input quantity m;
m	is a continuous variable;
N	is the number of input quantities X_m on which measurand Y depends.

Due to the functional relationship between the measurand Y and the input quantities, and in combination with an evaluation of measurement uncertainty, it is possible to obtain a quantitative statement of the measurement result and its corresponding uncertainty [67]. At this point, it has to be considered that the knowledge of the input quantities will always be insufficient and therefore has to be estimated in some cases. According to GUM [68], the insufficient knowledge about the distribution of the input quantities X_m can be described by an assignment of a probability density function (PDF) $g_{X_m}(\xi_m)$, whereby ξ_m are the possible values of the quantities which can be assumed. The expected value of the PDF is defined as:

$$x_m = \mathrm{E}[X_m] = \int_{-\infty}^{+\infty} g_{X_m}(\xi_m)\, \xi_m d\xi_m \tag{2.77}$$

The corresponding standard deviation of the expected value is the standard uncertainty u_{x_m}:

$$u_{x_m} = \left\{ \mathrm{E}\left[(X_m - x_m)^2 \right] \right\}^{1/2} = \left\{ \int_{-\infty}^{+\infty} g_{X_m}(\xi_m)(\xi_m - x_m)^2\, d\xi_m \right\}^{1/2} \tag{2.78}$$

where:

x_m	is the expected value of a quantity m;
$g_{X_m}(\xi_m)$	is the probability density function (PDF);
ξ_m	is a possible value of X_m;
u_{x_m}	is the standard uncertainty of the input estimate x_m.

The next step is to identify an appropriate PDF, which describes the existing knowledge of the quantity [67]. According to GUM, the estimation of the input quantities and their uncertainties is reduced to the methods Type A and Type B.

Type A

The Type A evaluation is used for input quantities X_m, which are estimated from n independent repeated observations $q_{m,k}$. The arithmetic mean or average $\overline{q_m}$ is then defined as:

$$\overline{q_m} = \frac{1}{n} \sum_{k=1}^{n} q_{m,k} \qquad (2.79)$$

With its standard deviation σ_{q_m}:

$$\sigma_{q_m} = \sqrt{\frac{1}{n-1} \sum_{k=1}^{N} \left(q_{m,k} - \overline{q_m}\right)^2} \qquad (2.80)$$

where:

$\overline{q_m}$	is the arithmetic mean or average of n independent repeated observations $q_{m,k}$;
n	is the number of independent repeated observations $q_{m,k}$;
$q_{m,k}$	is an independent observation;
k	is a continuous variable;
σ_{q_m}	is the standard deviation of n independent repeated observations $q_{m,k}$.

The expectation value x_m of the input quantity X_m is then defined as the mean value $\overline{q_m}$ (see Equation 2.79). And the corresponding standard uncertainty u_{x_m} of the input estimate x_m can be determined due to :

$$u_{x_m} = \frac{1}{\sqrt{n}} \sigma_{q_m} \qquad (2.81)$$

The corresponding PDF is a Gaussian distribution curve [67].

Type B

In this case, the estimate x_m of an input quantity X_m is not obtained from repeated observations. The corresponding uncertainty is evaluated using scientific judgement based on background information, such as, for example [68]:

- Data from previous measurements

- Knowledge of the behaviour and the properties of relevant materials and measurement devices

- Device specification from the manufacturer

- Information about the uncertainty provided in calibration certificates and other certificates

- Data about uncertainties assigned to reference data taken from handbooks

This evaluation of the standard uncertainty is called Type B.

If the estimate x_m is taken from a manufacturer's specification, calibration, certificate, handbook or other source, it is frequently stated that the value of X_m lies within the interval a_- to a_+ and its probability is equal to one for practical purpose. Otherwise, the probability that X_m lies beyond this specified range is zero. The corresponding PDF is rectangular. In this case the expectation value is the midpoint of the interval [67, 68]:

$$x_m = \frac{a_- + a_+}{2} \tag{2.82}$$

a_-, a_+ are the boundaries of the interval.

And the corresponding uncertainty is defined as [68, 67]:

$$u_{x_m} = \frac{1}{\sqrt{12}} \left(a_+ - a_-\right) \tag{2.83}$$

If the difference between the bounds a_+ and a_- can be denoted by $2a$, Equation 2.83 can be simplified [67, 68]:

$$u_{x_m} = \frac{a_+}{\sqrt{3}} = \frac{a_-}{\sqrt{3}} \tag{2.84}$$

In other cases, the information about the expanded uncertainty U is given in calibration certificates. In this case the expectated value x_m is equal to the estimate y of the measurand Y or the result of the measurement, respectively:

$$x_m = y \tag{2.85}$$

where:

y is the estimate of the measurand Y.

The uncertainty u_{x_m} can then be calculated using the following equation:

$$u_{x_m} = \frac{U_y}{k_p} \tag{2.86}$$

where:

U_y is the expanded uncertainty;

k_p is the coverage factor.

Furthermore, the coverage factor k_p should always be specified in the calibration certificate. After the evaluation of the relevant input quantities and their uncertainties, the combined uncertainty is calculated using the Gaussian uncertainty propagation:

$$u_y = \sqrt{\sum_{m=1}^{N} \left(\frac{\partial f}{\partial x_m}\right)^2 u_{x_m}^2 + \dots} \tag{2.87}$$

where:

u_y is the combined uncertainty.

However, Equation 2.87 is only valid, if the input quantities X_m are independent or uncorrelated. If some of the X_m are significantly correlated, the correlations have to be taken into account. The combined standard uncertainty is then defined as:

$$u_y = \sqrt{\sum_{m=1}^{N}\sum_{l=1}^{N} \frac{\partial f}{\partial x_m}\frac{\partial f}{\partial x_l} u\left(x_m, x_l\right)} = \sqrt{\sum_{m=1}^{N}\left(\frac{\partial f}{\partial x_m}\right)^2 u_{x_m}^2 + 2\sum_{m=1}^{N-1}\sum_{l=m+1}^{N} \frac{\partial f}{\partial x_m}\frac{\partial f}{\partial x_l} u\left(x_m, x_l\right)} \tag{2.88}$$

Moreover, the degree of the correlation between the expectation values x_m and x_l of the input quantities is characterized by the estimated correlation coefficient $r\left(x_m, x_l\right)$:

$$r\left(x_m, x_l\right) = \frac{u\left(x_m, x_l\right)}{u\left(x_m\right) u\left(x_l\right)} \tag{2.89}$$

With Equation 2.89, Equation 2.88 results in:

$$u_y = \sqrt{\sum_{m=1}^{N}\left(\frac{\partial f}{\partial x_m}\right)^2 u_{x_m}^2 + 2\sum_{m=1}^{N-1}\sum_{l=m+1}^{N} c_m c_l u_{x_m} u_{x_l} r\left(x_m, x_l\right)} \tag{2.90}$$

where:

$u\left(x_m, x_l\right)$ is the covariance of the input estimates x_m and x_l;

x_l is the expected value of a quantity l;

l is a continuous variable;

$r\left(x_m, x_l\right)$ is the correlation coefficient between the expectation values x_m and x_l;

c_m, c_l are the sensitivity coefficients of the expectation values x_m amd x_l of the input quantities.

The correlation coefficient $r\left(x_m, x_l\right)$ has a value between ± 1 and is zero if the input quantities are uncorrelated [67, 68].

Furthermore, the partial derivatives $\delta f/\delta x_m$ are equal to $\delta f/\delta X_m$ and are called sensitivity coefficients c_m. The sensitivity coefficient describes, how the output estimate y varies due to changes of the expectation values x_m of the input quantities. The component of the combined standard uncertainty $u_{m,y}$ is defined as:

$$u_{m,y} = c_m \cdot u_{x_m} \tag{2.91}$$

With the $u_{m,y}$ values, the cause and the contribution of the individual uncertainties compared to the total uncertainty of the measurement result can be identified [67, 68].

However, in some commercial, industrial and regulatory applications, it is usual and necessary to determine an uncertainty that defines an interval I_y about the measurement result y, as seen in Figure 2.35.

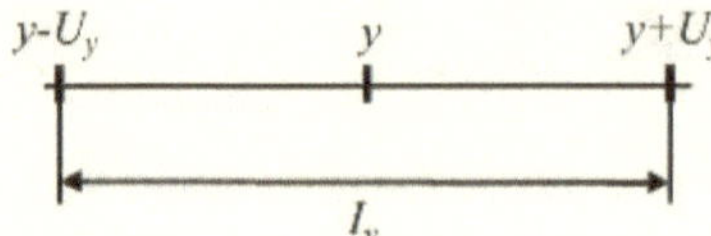

Figure 2.35: Expanded uncertainty U_y illustrated as an interval over the the measurement result y (own representation based on [67]).

The additional measure of the uncertainty is defined as expanded uncertainty U_y, which is obtained by multiplying the combined uncertainty u_y by a coverage factor k_p:

$$U_y = k_p \cdot u_y \tag{2.92}$$

The coverage factor k_p is chosen on the basis of the required level of confidence. For example, a coverage factor k_p with a value of 2 results in a level of confidence of approximately 95 %. To reach a level of confidence of nearly 99 %, a value of 3 is necessary.

The result of the measurement is then determined according to the following equation:

$$Y = y \pm U_y \tag{2.93}$$

In any case, the value of the coverage factor k_p or the level of confidence has to be stated together with the result of the measurement (see Equation 2.93) [67, 68].

CHAPTER 3

Methods

3.1 Used methods for road load determination

3.1.1 Coastdown method - road load determination on a proving ground

The test procedure for the CoastDown Method (CDM) is executed at the proving ground of the BMW Group in Aschheim. The test track consists of two parallel, straight lanes which are aligned $15°$ to the east-west axis. A schematic representation of the proving ground with the driving direction and its orientation is shown in Figure 3.1.

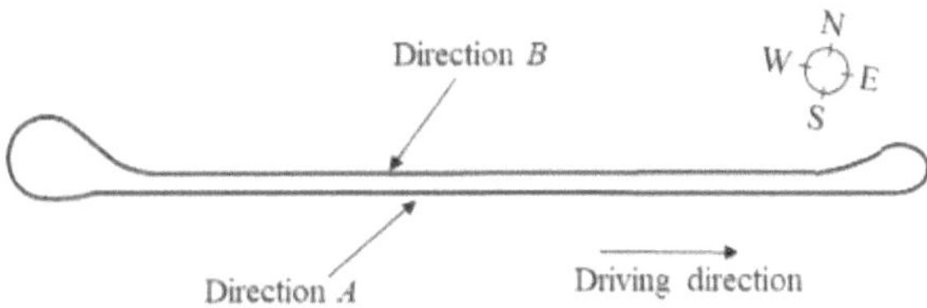

Figure 3.1: Schematic representation of the proving ground of the BMW Group in Aschheim.

The southern section has a length of 3,200 m and has a road surface of asphalt. With a length of 2,800 m, the northern lane has a road surface of concrete instead of asphalt on a section length of 2300 m.

The coastdown measurements are executed in a velocity range from $135\,\mathrm{km/h}$ to $15\,\mathrm{km/h}$ to determine the road load of a vehicle at the reference velocity points from $130\,\mathrm{km/h}$ to

$20\,\text{km/h}$ in incremental steps of $10\,\text{km/h}$. However, both the length of the northern and the southern lane are not long enough to carry out one coastdown run from $135\,\text{km/h}$ to $15\,\text{km/h}$. Instead, each coastdown run is split into runs from $130\,\text{km/h}$ to $75\,\text{km/h}$ and $75\,\text{km/h}$ to $15\,\text{km/h}$. Additionally, each coastdown run is carried out both in direction A and direction B to eliminate the influence of the wind and of the test track, as described in subsection 2.3.1 and in Figure 2.25.

During the test procedure, the vehicle velocity and time signals are recorded with a measurement device with a frequency of $10\,\text{Hz}$[10]. The weather data for air temperature, air pressure, wind velocity and wind direction which are necessary for such a road load correction to reference conditions are measured with a frequency of $1\,\text{Hz}$ at two stationary anemometers located at the test tracks (see Figure 2.24). The determined road load is corrected to reference conditions using Equations 2.45 to 2.49 (see in subsection 2.3.1).

3.1.2 Wind tunnel method and wind tunnel method extended

Wind tunnel method according to WLTP

According to the Wind Tunnel Method (WTM) described in subsection 2.3.2, the aerodynamic drag and the remaining components of the road load (mostly rolling resistance and drivetrain losses) are determined separately. The aerodynamic drag coefficient is measured in the BMW Group wind tunnel. The test section of the wind tunnel has a height of $13\,\text{m}$, a length of $18\,\text{m}$ and a width of $16\,\text{m}$ and includes a 5-belt system, comprising a centre belt with a length of $10\,\text{m}$ and a width of $1.1\,\text{m}$, and four wheel drive units. The vehicle is fixed by the rocker panels of the vehicle at a constant ride height. The ride height is equal to the ride height of the standing vehicle which is loaded with the vehicle test mass m_{test} defined for road load determination. The aerodynamic drag coefficient c_{w} is determined at wind and belt velocities of $140\,\text{km/h}$. In Figure 3.2 the BMW Group wind tunnel with the 5-belt system and a test vehicle (F46 216i) is shown. The force in the x direction ($F_{x,\text{Wind}}$) is measured with an accuracy of $0.97\,\text{N}$, a resolution of $0.1\,\text{N}$ and a repeatability of $0.39\,\text{N}$[11]. The aerodynamic drag coefficient c_{w} is then determined according to [69]:

$$c_{\text{w}} = \frac{2 \cdot F_{x,\text{Wind}}}{\rho_{\text{TS}} \cdot A_x \cdot U_{\text{TS}}^2} \tag{3.1}$$

with

$$U_{\text{TS}} = \sqrt{\frac{2 \cdot q_{\text{TS}}}{\rho_{\text{TS}}}} \tag{3.2}$$

and

$$q_{\text{TS}} = \frac{\kappa \cdot p_{\text{stat}}}{\kappa - 1} \cdot \left[\left(\frac{\Delta p}{p_{\text{stat}}} + 1 \right)^{\frac{\kappa-1}{\kappa}} - 1 \right] \tag{3.3}$$

---

[10]TBJ MESSTECHNIK. GPS - Velocity measurement system - gps10/CANID, 2019.
[11]MTS. Wind Tunnel Balance Calibration Report: Summary of One-site Balance Calibration, 2019.

Figure 3.2: The BMW Group wind tunnel with the 5-belt system (a: centre belt; b: wheel drive units) and the test vehicle F46 216i (adapted by permission from Isabell Vogeler: Different methods for road load determination in comparison: Wind tunnel, Wind tunnel method according to WLTP and Coastdown method [60], 2018).

Finally, the aerodynamic drag of the vehicle $F_{j,\mathrm{Air}}$ is then calculated according to:

$$F_{j,\mathrm{Air}} = c_\mathrm{w} \cdot A_x \cdot \frac{\rho_0}{2} \cdot v_j^2 \tag{3.4}$$

where:

c_w	is the aerodynamic drag coefficient (see Equation 3.1);
$F_{x,\mathrm{Wind}}$	is the measured force of the aerodynamic drag in x direction in N;
ρ_TS	is the air density in the test section in $\mathrm{kg/m^3}$;
A_x	is the frontal area of the vehicle in $\mathrm{m^2}$;
U_TS	is the air velocity in the test section of the wind tunnel in $\mathrm{m/s}$ (see Equation 3.2);
q_TS	is the dynamic pressure in the test section of the wind tunnel in Pa (see Equation 3.3);
κ	is the ratio of the specific heats for air and has a value of 1.4;
p_stat	is the absolute static pressure measured in the plenum of the wind tunnel in Pa;
Δp	is the differential pressure between the total and static pressure port in Pa;
$F_{j,\mathrm{Air}}$	is the aerodynamic drag at the reference velocity point v_j in N;
ρ_0	is the dry air density, which has a value of 1.189 $\mathrm{kg/m^3}$;
v_j	is the reference velocity point in $\mathrm{m/s}$.

The differential pressure is measured with a PDT (Difference Pressure Transducer) between the settling chamber and the plenum. Additionally, the static pressure is related to the total pressure, which is necessary for the calculation of the air velocity. The static pressure is determined by a PT (Pressure Transducer). In Figure 3.3 the described pressure measurement configuration in the wind tunnel is schematically illustrated.

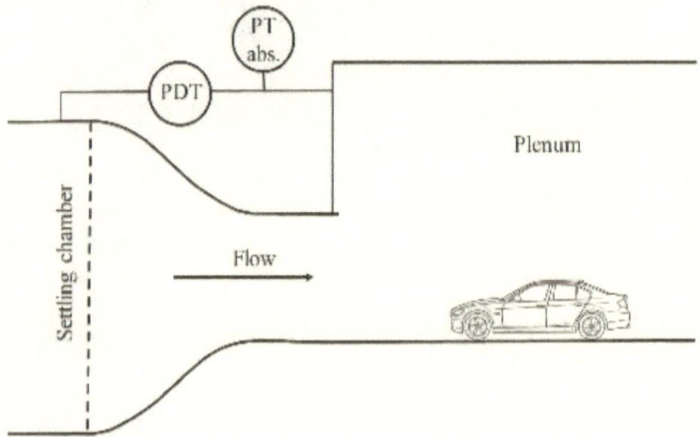

Figure 3.3: Schematic representation of the pressure measurement configuration in the BMW Group wind tunnel (own representation based on [69]).

The remaining components of the road load (mostly rolling resistance and drivetrain losses) are determined using the flat belt dynamometer of the BMW Group, which is schematically illustrated in Figure 3.4 and shown in Figure 3.5.
The test vehicle stands with each wheel on a wheel drive unit (WDU), which exists of a drive unit with an integrated engine and a second roller. Over these two rollers a flat belt, laminated with a Safety Walk^{TM} coating, is tensioned. To ensure that the vehicle can roll on the flat belt, an air bearing is located directly under the tyre contact patch and the flat belt. As a result, both the Safety Walk^{TM} coating on the flat belt and the air bearing provide a rolling characteristic of the vehicle wheels which is similiar to the rolling characteristic on a real road. Additionally, each WDU is supported by further air bearings, which allow an exact force measurement in x direction with an accuracy of $0.3\,N$ for each load cell[12]. A cooling fan stands in front of the vehicle to cool the tyres and the running engine of the vehicle. Using this configuration it is ensured that the measurement results do not contain aerodynamic forces (compare the results in subsection 4.1.1). The vehicle is usually fixed by a restraint system which uses the tow hooks at the front and at the rear of the vehicle. If the restraint system is mounted horizontally to the vehicle, only the rolling resistance and drivetrain losses are measured (see subsection A.3 in the appendix). The temperature of the test cell is $23\,°C$. The test procedure to determine the road load of a vehicle and the correction terms are taken from the description in subsection 2.3.2.

---

[12] AERODYNAMISCHES VERSUCHSZENTRUM BMW GROUP. Flachbahnprüfstand Verifikation Kraftmesseinrichtung: Antriebseinheit FL, Antriebseinheit FR, Antriebseinheit HL, Antriebseinheit HR, 2020.

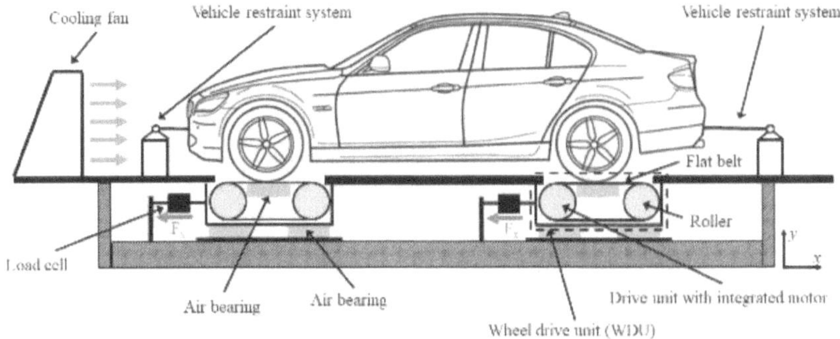

Figure 3.4: Schematic representation of the flat belt dynamometer of the BMW Group (adapted by permission from Springer Nature Customer Service Centre GmbH: Springer Nature, 18. Internationales Stuttgarter Symposium [8], 2018).

Figure 3.5: The flat belt dynamometer of the BMW Group with a test vehicle (adapted by permission from Isabell Vogeler: Different methods for road load determination in comparison: Wind tunnel, Wind tunnel method according to WLTP and Coastdown method [60], 2018).

Wind tunnel method extended

As already stated in subsection 2.2.2, the road/belt surface texture has a significant influence on the absolute value of the rolling resistance. Moreover, in Günter [38], the rolling resistance of a tyre with different wheel loads is on average 20 % to 30 % higher on asphalt when compared to a surface coated with Safety Walk^{TM}. These effects are not considered in the correction terms according to GTR No. 15. Therefore, in the present study a correction method is developed which supplements the correction terms according to GTR No. 15 (see subsection 2.3.2). In Figure 4.45 it can be seen that the ratio between the rolling resistance coefficient on a Safety Walk^{TM} surface and on a steel surface is nearly constant over the investigated velocity range from 20 ^{km}/$_{\text{h}}$ to 130 ^{km}/$_{\text{h}}$. For the ratio of the rolling resistance coefficient between a Safety Walk^{TM} surface and an asphalt surface the same behaviour is assumed. Therefore, a correction factor for the rolling resistance part is introduced in the following.

According to this correction method, the correction factor K_{ext} is applied to the rolling resistance part $F_{j,\text{Roll}}$ of the corrected flat belt dynamometer result $F_{j,\text{Dyno}}$ according to GTR No. 15. (see Equation 2.53):

$$F_{j,\text{WTMext}} = (F_{j,\text{Dyno}} - F_{j,\text{Drive}}) \cdot K_{\text{ext}} + F_{j,\text{Drive}} + F_{j,\text{Air}}$$
$$= F_{j,\text{Roll}} \cdot K_{\text{ext}} + F_{j,\text{Drive}} + F_{j,\text{Air}} \quad (3.5)$$

where:

$F_{j,\text{WTMext}}$	is the road load of a vehicle according to the wind tunnel method extended in combination with the additional correction factor K_{ext} in N (see Equation 2.53);
$F_{j,\text{Dyno}}$	is the corrected road load measured at the flat belt dynamometer at the reference velocity point v_j in N;
$F_{j,\text{Roll}}$	is the rolling resistance of a vehicle at the reference velocity point v_j and part of the vehicle road load measured at the flat belt dynamometer $F_{j,\text{Dyno}}$ in N;
K_{ext}	is the constant correction factor according to the wind tunnel method extended;
$F_{j,\text{Drive}}$	are the drivetrain losses of a vehicle at the reference velocity point v_j and part of the vehicle road load measured at the flat belt dynamometer $F_{j,\text{Dyno}}$ in N;
$F_{j,\text{Air}}$	is the aerodynamic drag at the reference velocity point v_j in N.

For the separation of the road load measured at the flat belt dynamometer into its components rolling resistance $F_{j,\text{Roll}}$ and drivetrain losses $F_{j,\text{Drive}}$, the torque meter measurement method is used, which is explained in section 3.3.4.

Similar to the coastdown method and the wind tunnel method, the road load coefficients

f_0^c, f_1^c and f_2^c are determined with a polynominal regression of second order and the road load $F_{j,\text{WTMext}}^c$ can then be calculated on the basis of the road load coefficients equivalent to Equation 2.2. This method is called Wind Tunnel Method extended (WTMext) in the following.

As the wind tunnel method, the wind tunnel method extended is compared with the coastdown method using the difference in cycle energy demand:

$$\epsilon_{\text{WTMext}} = \frac{E_{\text{WTMext}}}{E_{\text{CDM}}} - 1 \tag{3.6}$$

where:

ϵ_{WTMext}	is the difference in cycle energy demand;
E_{WTMext}	is the total cycle energy demand over a complete test cycle with the road load determined using the wind tunnel method extended (WTMext) in J (for the calculation see Equation 2.66);
E_{CDM}	is the total cycle energy demand over a complete test cycle with the road load determined using the coastdown method (CDM) in J (for the calculation see Equation 2.66).

The total cycle energy demands E_{WTMext} and E_{CDM} are calculated using the equations given in subsection 2.4.1. The determination of the additional correction factor K_{ext} and the evaluation of this method are described in section 4.2.

Note: The wind tunnel method extended is different to the wind tunnel method described in GTR No. 15. The wind tunnel method extended is only used in the context of the present study.

3.1.3 AEROLAB method

The main part of this **book** investigates whether it is possible to determine the road load of a vehicle containing the aerodynamic drag, the rolling resistance and the drivetrain losses in a wind tunnel. The AEROLAB wind tunnel of the BMW Group provides such new possibilities.

The AEROLAB wind tunnel consists of a single-belt-rolling-road system, on which the vehicle is positioned. Additionally, the vehicle is fixed by the front wheel hubs via a vehicle fixation system. The aerodynamic drag of a vehicle is determined with a Horizontal Force Measurement System (HFMS). It is schematically shown in Figure 3.6.
Both the single-belt-rolling-road system and the vehicle fixation system are located on a weighing plate. This weighing plate is floating without contact to the basement ground on an oil film from six hydrostatic bearings. The aerodynamic forces are measured with four load cells with two pairs measuring the forces in x and y directions, respectively.

Due to this construction, only the aerodynamic drag is measured. Additionally, due to the vehicle fixation system at the wheel hubs, the aerodynamic forces are measured depending on the ride height, which itself depends on the wind and the vehicle velocity. However, the rolling resistance and drivetrain losses of a vehicle are not included, because they are internal forces of the vehicle [70, 71].

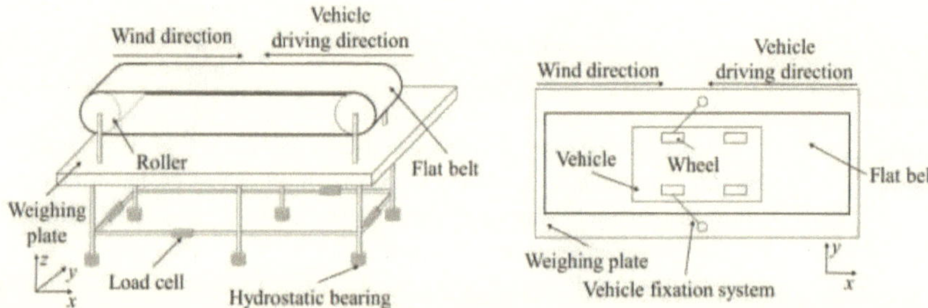

Figure 3.6: Schematic representation of the horizontal force measurement system of the AEROLAB wind tunnel of the BMW Group (adapted by permission from Springer Nature Customer Service Centre GmbH: Springer Nature, 18. Internationales Stuttgarter Symposium [8], 2018).

To measure both the aerodynamic drag and the internal forces, two separate load cells are integrated into the vehicle fixation system, as shown in Figure 3.7 and Figure 3.8.
Both load cells are able to measure a maximal force of $5500\,\mathrm{N}$ in the x and y directions. The accuracy is $0.6\,\mathrm{N}$ according to manufacturer's specification of MTS[13]. The surface of the single-road-rolling system is steel.
For the determination of the road load, a test procedure similiar to test procedures according to WLTP for the wind tunnel method is chosen. However, there are several limitations which have to be taken into account:

- The wind tunnel has no exhaust extraction system. Therefore, it is not possible to measure with a running engine.

- As the wind tunnel is not able to do a driving simulation, the acceleration and the braking phase at the beginning of the test procedure (see Figure 2.28) are not possible.

- The stabilization time of $4\,\mathrm{seconds}$ prior to each measurement point used at the flat belt dynamometer is not sufficient. The wind in the wind tunnel needs more time to stably reach every reference velocity point.

---

[13]MTS. Specification of the load cells: Model 670.67B-10, 2017.

Figure 3.7: The AEROLAB wind tunnel with the single-belt-rolling-road system (a), the vehicle fixation system (b) and the test vehicle F33 430i (adapted by permission from Springer Nature Customer Service Centre GmbH: Springer Nature, 18. Internationales Stuttgarter Symposium [8], 2018).

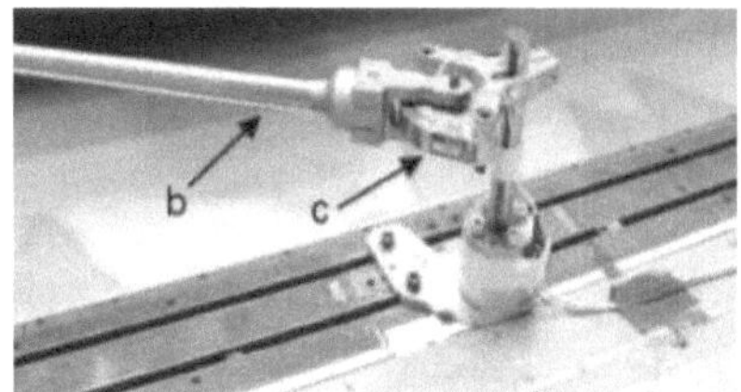

Figure 3.8: The vehicle fixation system (b) with integrated load cell (c) (adapted by permission from Springer Nature Customer Service Centre GmbH: Springer Nature, 18. Internationales Stuttgarter Symposium [8], 2018).

For these reasons, a test procedure is developed, which is different from the original test procedure defined in GTR No. 15, and has the following features:

- During the complete test procedure the engine is off.

- For the warm-up phase a velocity of $144\,\mathrm{km/h}$ is chosen.

- The braking phase is omitted. Instead, it is ensured that there are no residual brake forces in the measurement results.

- The missing residual brake forces have to be determined with an alternative method.

- The stabilization time of 4 seconds is increased up to 20 seconds for the reference velocity points from $130\,\mathrm{km/h}$ to $30\,\mathrm{km/h}$ and to 30 seconds for the reference velocity point of $20\,\mathrm{km/h}$.

The modified test procedure without the braking phase (woB), with alternative warm-up phase (ALT), and with the measurement phase with stabilized velocity (SV) is shown in Figure 3.9. The resulting losses measured in the AEROLAB wind tunnel without the residual brake forces are denoted $F_{j,\mathrm{WT,woB}}$. These losses are corrected to reference conditions using Equation 2.45, corresponding to the coastdown method. Only the correction factor w_1 is set to zero, as there is no headwind and tailwind in the wind tunnel.

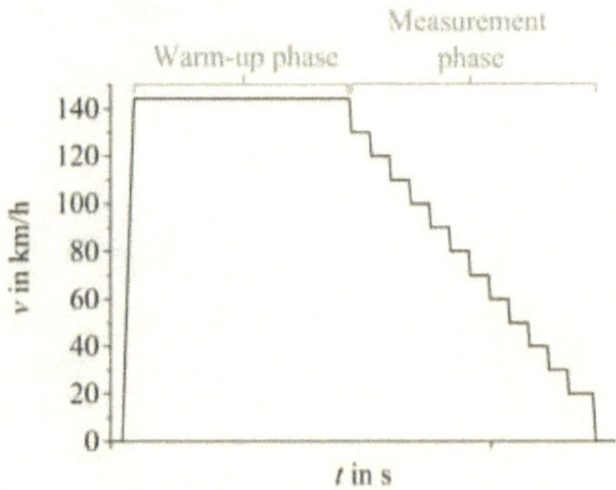

Figure 3.9: Modified test procedure without braking phase SV ALT woB with an increased stabilization time of 20 seconds for the reference velocity points ranging from $130\,\mathrm{km/h}$ to $30\,\mathrm{km/h}$ and of 30 seconds for the reference velocity point $20\,\mathrm{km/h}$ prior to each measurement point.

To determine the missing residual brake forces, an alternative method is developed, which is implemented on the flat belt dynamometer. At first, the losses of a vehicle $F_{j,\mathrm{Dyno}}$ using the test procedure SV ALT wB according to WLTP for the wind tunnel method, as shown in Figure 2.28, are determined at the flat belt dynamomter. Afterwards, the losses without residual brake forces $F_{j,\mathrm{Dyno,woB}}$ are measured using the modified test procedure SV ALT woB (see Figure A.1). In this case, the stabilization time is set to 4 seconds and the engine is running, as the test procedure is now conducted on the flat belt dynamomter. The resulting residual brake forces $F_{j,\mathrm{Brake}}$ can then be calculated using the following equation:

$$F_{j,\mathrm{Brake}} = F_{j,\mathrm{Dyno}} - F_{j,\mathrm{Dyno,woB}} \tag{3.7}$$

where:

$F_{j,\mathrm{Brake}}$	is the residual brake force at the reference velocity point v_j in N;
$F_{j,\mathrm{Dyno}}$	is the corrected road load of a vehicle at the reference velocity point v_j to reference conditions determined using the wind tunnel method in N;
$F_{j,\mathrm{Dyno,woB}}$	is the corrected road load of a vehicle without residual brake forces measured at the flat belt dynamometer in N.

The resulting road load of a vehicle $F_{j,\mathrm{AM}}$ according to the above described AEROLAB Method (AM) is the sum of the road load without residual brake forces measured in the AEROLAB wind tunnel and the separately determined residual brake forces $F_{j,\mathrm{Brake}}$ on the flat belt dynamometer:

$$F_{j,\mathrm{AM}} = F_{j,\mathrm{Brake}} + F_{j,\mathrm{WT,woB}} \tag{3.8}$$

where:

$F_{j,\mathrm{AM}}$ is the road load of a vehicle at the reference velocity point v_j determined using the AEROLAB method in N;

$F_{j,\mathrm{WT,woB}}$ is the road load of a vehicle without residual brake forces measured in the AEROLAB wind tunnel in N.

Similar to the previous described methods, the road load coefficients f_0^c, f_1^c and f_2^c are determined with a polynominal regression of second order and the road load $F_{j,\mathrm{AM}}^c$ can be calculated on the basis of the road load coefficients equivalent to Equation 2.2. The above described AEROLAB method is schematically illustrated in Figure 3.10 for a better understanding.

Equivalent to the wind tunnel method, the AEROLAB method is compared with the coastdown method according to the difference in cycle energy demand:

$$\epsilon_{\mathrm{AM}} = \frac{E_{\mathrm{AM}}}{E_{\mathrm{CDM}}} - 1 \tag{3.9}$$

where:

ϵ_{AM} is the difference in cycle energy demand;

E_{AM} is the total cycle energy demand over a complete test cycle with the road load determined using the AEROLAB method (AM) in J (for the calculation see Equation 2.66);

E_{CDM} is the total cycle energy demand over a complete test cycle with the road load determined using the coastdown method (CDM) in J (for the calculation see Equation 2.66).

The total cycle energy demand can be calculated using the equations given in subsection 2.4.1.

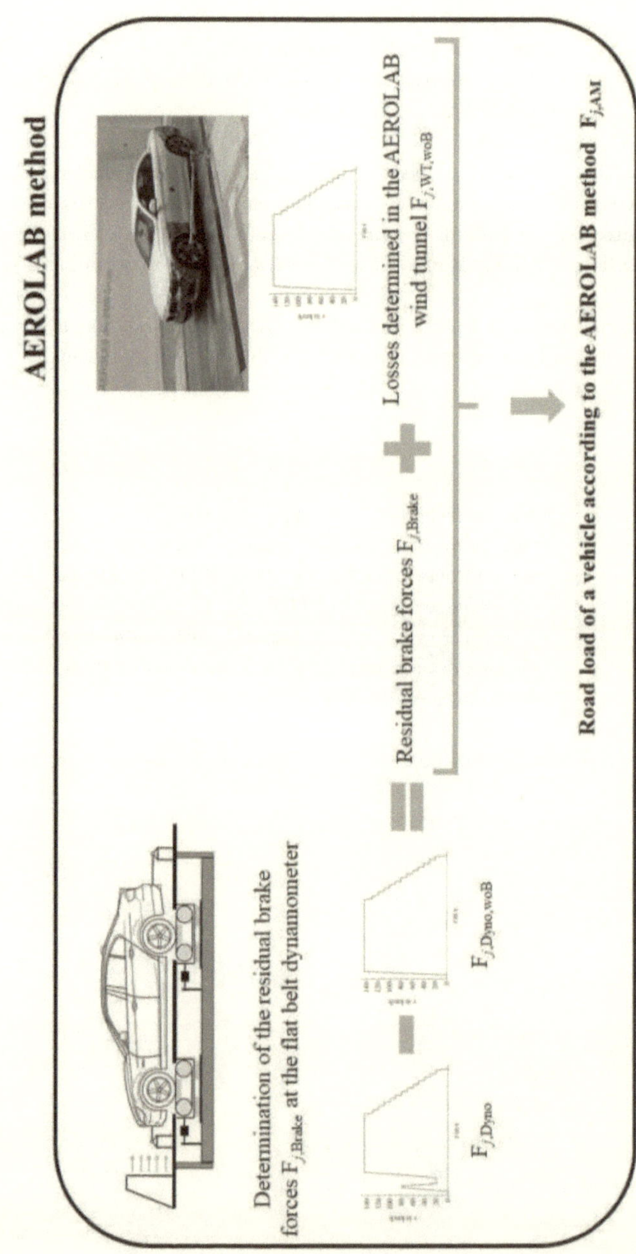

Figure 3.10: Schematic representation of the AEROLAB method.

3.2 Test vehicles

For the main part of this study, a BMW F46 216i with front wheel drive is used as a test vehicle (see in Figure 3.2). Additionally, for the basic research at the flat belt dynamometer, a BMW F33 430i with rear wheel drive (see Figure 3.5) is used. Furthermore, for the verification of the AEROLAB method, a BMW G30 518d with rear wheel drive and a MINI F60 Countryman with front wheel drive are chosen. All these vehicles have a manual transmission. Further information on these vehicles and about the vehicles, which are used to develop and to verify the wind tunnel method extended, is provided in Table A.1 and in Table A.2 in the appendix.

3.3 Further measurement technology

For a detailed comparison of the road load determined with the different methods (coast-down method, wind tunnel method extended and AEROLAB method) the vehicles are partially equipped with additional measurement technology. These measurement systems are introduced in the next subsections. All sensor data is recorded with the INCA 7.2 software.

3.3.1 Temperature measurement of transmission oils and tyre treads

The oil temperature of the manual transmissions and the oil temperature of rear axle differentials for vehicles with rear axle drive are measured by temperature sensors in the oil drain plugs. Furthermore, the tyre temperatures are monitored during the measurements. However, as the temperatures of different tyre parts (for example bead, shoulder, tread ...) differ considerably, it is difficult to obtain representative values [42]. Therefore, infrared sensors are mounted in the wheel housings of the test vehicle F46 216i, to monitor the tyre tread temperatures. In this way, it is ensured that on the street and at the test benches the temperatures are always measured at the same positions. The sensors have an accuracy of $1\,°C$ and a resolution of $0.1\,°C$ in the temperature range from -40 °C to at least +600 °C[14]. The positions of the sensors in the wheel housing are marked with a yellow ellipse in Figure 3.11, as an example for the front wheel, and in Figure 3.12, as an example for the rear wheel.

However, when the torque meter measurement method (see chapter 3.3.4) is used, these infrared sensors can not be used, as their measurement point is relocated due to the additional measurement technology between the wheel hub and the rim (see Figure 3.14). Instead, other infrared sensors, which are installed on the ground of the flat belt dynamometer in front of each wheel, are used (see the marked positions in Figure 3.13). Their measurement point can be adjusted exactly to the tyre treads. They have a measurement accuracy of $\pm 1\,°C$[15].

---

[14]OPTRIS. Bedinungsanleitung: IR-Sensor CT, 2005.

[15]RAYTEK - A FLUKE COMPANY. Datenblatt: Raytek Compact Serie, 2009.

Figure 3.11: Infrared sensors for the tyre tread temperature (yellow ellipse) and for the laser sensor for ride height change measurement mounted in front of the front wheels (red ellipse).

Figure 3.12: Infrared sensors for the tyre tread temperature (yellow ellipse) and for the laser sensors for ride height change measurement mounted behind the rear wheels (red ellipse).

Figure 3.13: Infrared sensors installed on the ground of the flat belt dynamometer to measure the tyre tread temperature (adapted by permission from Isabell Vogeler: Different methods for road load determination in comparison: Wind tunnel, Wind tunnel method according to WLTP and Coastdown method [60], 2018).

3.3.2 Tyre air temperature and tyre pressure measurement

The air temperature and the air pressure of the tyres are recorded during the measurements via the Tire Pressure Monitoring System (TPMS), which is usually installed in vehicles of the BMW Group. In the vehicles BMW F46 216i and MINI F60 Countryman a system from Continental is installed. The maximum measurement error of the air temperature for this system is $\pm 3\,°C$ over a temperature range from about -20 °C to +70 °C. The measurement error of the tyre pressure is 0.1 bar in the pressure ranges from 1 bar to 9 bar and from 0 °C to +70 °C, respectively [16]. In contrast, in the vehicle BMW G30 518d, a system from Sensata is used. Here, the measurement error of the tyre pressure is ± 0.075 bar in the temperature range from 0 °C to +50 °C[16].

3.3.3 Ride height change

The ride height change of a vehicle is measured with four OMRON ZX-LD100 laser sensors with a measurement range of 100 ± 40 mm and a resolution of 16 µm. They are mounted in the vehicle underbody in front of the front wheels and behind the rear wheels respectively, as shown in the marked area in the Figures 3.11 and 3.12 with a red ellipse.

The sensor data are recorded via the vehicle CAN data bus with the INCA 7.2 software. At this point it is clarified that the measurement results of the laser sensors are not equal to the ride height. Instead, the ride height at each wheel is determined with the aid of equations of planes, which are set up with the measurement points of the laser sensors. In addition to that, the signals measured during the coastdown runs on a proving ground are initially smoothed with a moving average filter. Afterwards, the arithmetic averaged ride height of six coastdown runs for each direction A and B and each wheel is calculated separately.

3.3.4 Torque meter measurement method for separation of rolling resistance and drivetrain losses at the flat belt dynamometer

As stated in subsection 3.1.2, the measurement results $F_{j,\text{Dyno}}$ from the flat belt dynamometer are the sum of rolling resistance $F_{j,\text{Roll}}$ and drivetrain losses $F_{j,\text{Drive}}$ of a vehicle. However, to investigate for example the influence of different ambient temperatures and flat belt surfaces, a measurement method is developed which enables the measurement result of the flat belt dynamometer to be separated into its two components [72].

In subsection 2.2.3 some methods are already mentioned, which allow the measurement of the drivetrain losses. The first possibility is to use wheel torque transducers. However, the disadvantage of this method is that the transducer replaces the middle part of the rims. Therefore, the original rims cannot be used during the test. Additionally, for each wheel size a separate set of transducers is necessary [54, 73]. Another option is to use strain gauges

---

[16]BMW AG - TYRE PRESSURE MONITORING SYSTEMS. Pressure and temperature accuracy of the Tire Pressure Monitoring System, e-mail, 17.07.2019.

which are mounted on the output shafts of the vehicle. However, the application of strain gauges is very expensive, as can be seen from the required steps for the commissioning described in [55]. As a consequence, this method requires increased time to put this measuring technology in operation and is therefore not a good choice for quick testing of many different vehicles and drivetrains [73].

As a result, a new measurement method is developed by Untermaierhofer, Petz and Vogeler [72] to separate the rolling resistance and the drivetrain losses. This custom-built torque flange is explained in detail in the following and is shown in Figure 3.14.

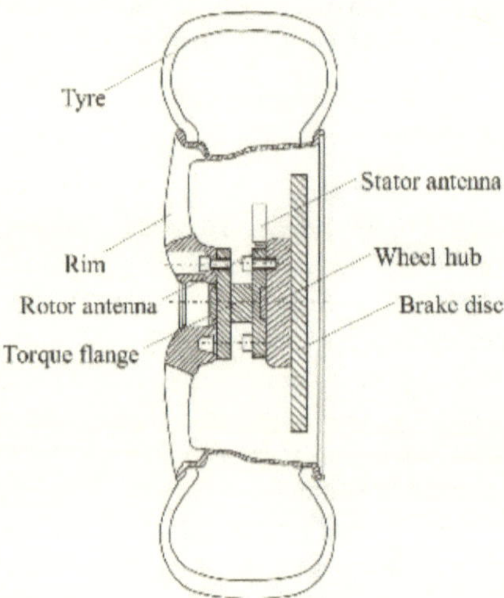

Figure 3.14: Schematic representation of the torque flange, which is installed between wheel hub and rim (adapted by permission from Rainer Untermaierhofer, Verfahren zum Erfassen eines auf ein Rad eines Kraftfahrzeugs wirkenden Drehmoments, Drehmomenterfassungsvorrichtung sowie Prüfstand [72], 2020).

The torque meter (hereafter called TOM) is a custom-built torque flange, which is mounted between the wheel hub and the rim of a vehicle. The data is transmitted via a telemetry system, which consists of a rotor and a stator antenna. The rotor antenna is installed on the torque flange and transfers the data via induction to the stator antenna which is fixed on the vehicle body. Furthermore, it has to be considered that the track of the vehicle is widened by the width of the flange by about 63 mm at each side [72].

To determine the drivetrain losses, including also the wheel hub losses for each wheel separately and at the reference velocity point v_j, the following test procedures can be used:

- Test procedure without braking phase (see for example Figure A.1)

- Test procedures with braking phase (see Figures 2.27 or 2.28): Then, the determined drivetrain losses contain not only the wheel hub losses but also residual brake forces.

To obtain a wheel specific force $F_{j,\text{TOM,Wheel}}$, the measured torque $M_{j,\text{TOM,Wheel}}$ can be converted with the following equation [72]:

$$F_{j,\text{TOM,Wheel}} = \frac{M_{j,\text{TOM,Wheel}}}{r_{\text{dyn}}} = F_{j,\text{Drive,Wheel}} \qquad (3.10)$$

where:

$F_{j,\text{TOM,Wheel}}$	is the wheel specific force determined with TOM at the reference velocity point v_j in N;
$M_{j,\text{TOM,Wheel}}$	is the wheel specific torque measured with TOM at the reference velocity point v_j in Nm;
r_{dyn}	is the dynamic rolling radius in m;
$F_{j,\text{Drive,Wheel}}$	are the wheel specific drivetrain losses at the reference velocity point v_j in N.

In this case, the wheel specific force $F_{j,\text{TOM,Wheel}}$, determined with the measured torque $M_{j,\text{TOM,Wheel}}$, is equal to the wheel specific drivetrain losses $F_{j,\text{Drive,Wheel}}$ of the vehicle [72]. The sum of all four $F_{j,\text{Drive,Wheel}}$ of a vehicle is then equal to the drivetrain losses $F_{j,\text{Drive}}$ (including the wheel hub losses) of a vehicle. The rolling resistance $F_{j,\text{Roll}}$ can then be calculated using the following equation:

$$F_{j,\text{Roll}} = F_{j,\text{Dyno}} - F_{j,\text{Drive}} \qquad (3.11)$$

If the vehicle is driven by its own engine during the measurement phase, instead by the test bench, the rolling resistance $F_{j,\text{Roll}}$ is measured with TOM instead of the drivetrain losses $F_{j,\text{Drive}}$.

The dynamic rolling radius r_{dyn} is calculated from the dynamic rolling circumference U_{dyn} of the wheel, which is defined as the distance of one full revolution of the wheel at a velocity of $60\,\text{km/h}$, according to DIN 70020. The values can also be looked up in tables from the tyre manufacturers [16].

In Table 3.1 the tyre sizes of the vehicles introduced in section 3.2 of this chapter are listed. Additionally, the dynamic circumference U_{dyn} (taken from [74]) and the resulting dynamic rolling radius r_{dyn} are given. These values are used in the following to calculate the wheel specific force $F_{j,\text{TOM,Wheel}}$ from the measured torque $M_{j,\text{TOM,Wheel}}$.

For each torque meter with a measuring range from $0\,\text{Nm}$ to $80\,\text{Nm}$, the maximum characteristic curve deviation is $0.084\,\%$[17] [73]. Assuming a dynamic radius of $320\,\text{mm}$, this results in a deviation of $0.21\,\text{N}$.

---

[17]MANNER SENSORTELEMETRIE. Werkskalibrierschein: Drehmomentaufnehmer 79999, 2018.

Table 3.1: Tyres sizes of the mainly used test vehicles

Vehicle	Tyre size				U_{dyn} in mm	r_{dyn} in mm
	Nominal section width in mm	Nominal Aspect Ratio	Construction Code	Nominal rim diameter in in		
F46 216i	225	45	R	18	2010 [74]	320
F33 430i	225	45	R	18	2010 [74]	320
G30 518d	225	55	R	17	2074 [74]	330
F60 Countryman	225	50	R	18	2083 [74]	332

CHAPTER 4

Results and discussion

4.1 Flat belt dynamometer

In this section, the road load determination at the flat belt dynamometer is investigated in detail. The topics will be:

- Influence of the cooling fan airstream on the resulting force measurement

- Road load difference due to the different test procedures stated in subsection 2.3.2

- Introduction of the torque meter measurement method, which enables the flat belt dynamometer results to be separated into its components drivetrain losses and rolling resistance (see subsection 3.3.4)

- Influence of the tyre inflation pressure

- Influence of the vehicle position on the flat belt

- Influence of the ambient temperature

Note: In this section the presented results for the road load determined at the flat belt dynamometer do not contain aerodynamic drag forces.

4.1.1 Influence of the cooling fan airstream on the force measurement

As already described in subsection 3.1.2, the rolling resistance and the drivetrain losses of a vehicle are determined using the flat belt dynamometer. The cooling fan in front of the vehicle is only used to cool the tyres and the running engine during the measurements. It does not attempt to simulate the airflow for the aerodynamic drag.

To demonstrate that the airstream of the cooling fan has no influence on the force measurement at all four Wheel Drive Units (WDUs), measurements with the test procedure SV ALT woB (see Figure A.1) but without a vehicle on the test bench are executed. The test procedure consists of the alternative warm-up phase (ALT) and of the measurement phase with stabilized velocity (SV). The braking phase at the beginning is omitted (woB). In Figure 4.1 the averaged resulting forces $F_{j,x}$ in x direction at each WDU front left FL (black line with squares), front right FR (dark grey line with dots), rear left RL (grey line with upward triangles) and rear right RR (light grey line with downward triangles) are plotted over the reference velocity points v_j. Additionally, the limit due to the accuracy of ± 0.3 N of each load cell [12] is illustrated as dashed black lines.

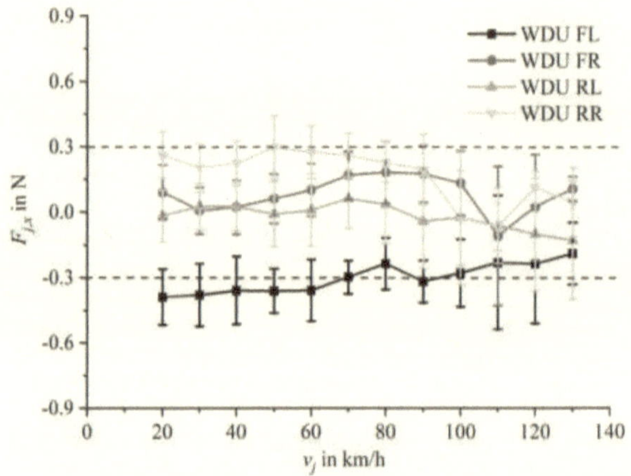

Figure 4.1: Averaged resulting forces $F_{j,x}$ in x direction at each WDU front left FL (black line with squares), front right FR (dark grey line with dots), rear left RL (grey line with upward triangles) and rear right RR (light grey line with downward triangles) of measurements at the flat belt dynamometer using the test procedure SV ALT woB but without vehicle.

It can be seen that the measured forces in the x direction are mainly within the limits of ± 0.3 N, which indicates that the forces induced by the airstream of the cooling fan cannot be resolved by the load cells. The maximum averaged force is about -0.4 N at the reference velocity point of 20 km/h for the WDU FL.

In the next step, it is investigated whether the airstream of the cooling fan has an effect on the measured flat belt dynamometer results of a vehicle. For this purpose, the vehicle is driven by the test bench with a constant velocity of 144 km/h (black dotted line) and with the running cooling fan at a velocity of 130 km/h (black dashed line). After the tyre tread temperatures at the tyre front left (FL, grey line) and front right (FR, dark grey line) have reached a constant level, the cooling fan is turned off. In Figure 4.2 the influence on the measured forces at the front axles (black line) is illustrated. To evaluate the influence, the force at the front axle is averaged for 30 seconds before and after the cooling fan is turned off. Both time ranges are marked as grey bars. The difference between the two averaged force values is about 0.7 N. Considering that the load cells of the front axles have in total an accuracy of $\pm 0.6\,N$[12], the difference is below the threshold of measurability. Furthermore, an increase of the force is observed, at the moment the fan is turned off, although a decrease was expected due to missing additional aerodynamic forces caused by the cooling fan airstream. This phenomenon can be explained by the decrease of the tyre tread temperatures at the moment the cooling fan is turned off. The influence of the tyre tread temperature on the road load measured at the flat belt dynamometer is further investigated in subsection 4.1.6.

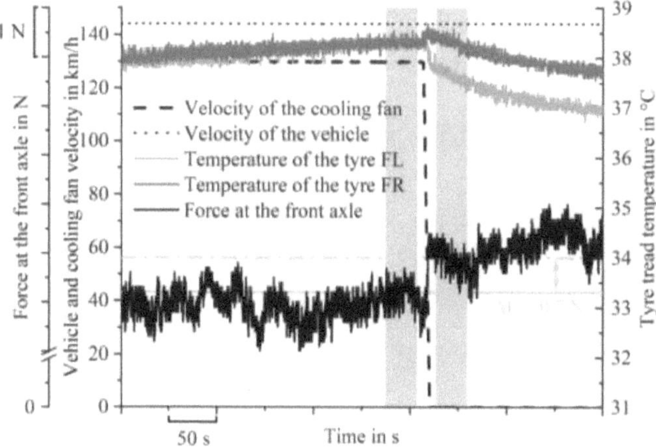

Figure 4.2: Vehicle velocity (black dotted line), cooling fan velocity (black dashed line), tyre tread temperature at the tyre front left (FL, grey line), at the tyre front right (FR, dark grey line) and the force at the front axle (black line) over time.

In Figure 4.3 the same procedure is plotted for the rear axle. The corresponding difference of the averaged force, before and after the cooling fan is turned off, is only 0.3 N. The comparatively lower difference can be explained by the the tyre tread temperatures of the rear tyres, which remain almost constant over the investigated period of time. This shows that the cooling effect of the fan is lower for the rear axle than for the front axle.

Concluding these results, the airstream of the cooling fan causes no aerodynamic forces influencing the flat belt dynamometer measurements. Moreover, only the influence at a vehicle velocity of 144 $^{km}/_h$ is investigated, as aerodynamic drag forces are decreasing with the square of the decreasing velocity. Therefore, a further investigation at the reference velocity points v_j in the range from 130 $^{km}/_h$ to 20 $^{km}/_h$ is left out.

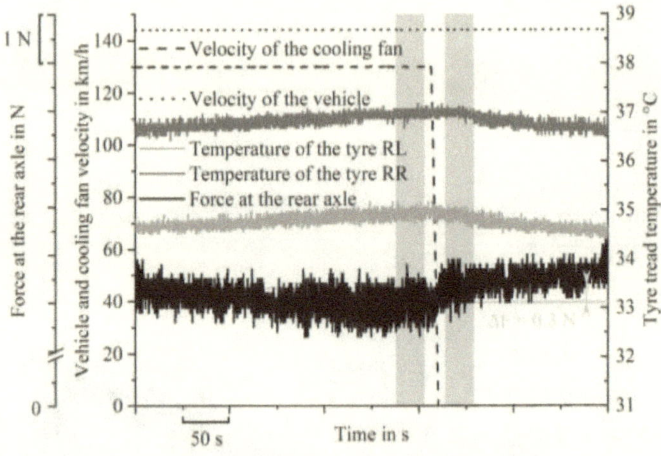

Figure 4.3: Vehicle velocity (black dotted line), cooling fan velocity (black dashed line), tyre tread temperature at the tyre front left (RL, grey line), at the tyre front right (RR, dark grey line) and the force at the rear axle (black line) over time.

4.1.2 Differences in road load due to the chosen test procedure

For the road load determination with the Wind Tunnel Method (WTM) according to GTR No. 15, four different test procedures are allowed. The different procedures have already been introduced in subsection 2.3.2 and are discussed in the following. For this discussion the test vehicle BMW F33 430i (see subsection 3.2) is chosen. This vehicle has a rear wheel drive and a manual transmission.

First, the test procedures, which simulate the CoastDown Method (CDM) on a test bench, are investigated. In Figure 4.4 the determined road load between the measurement phase by deceleration with the standard warm-up phase CD STD (grey line with dots) and with the alternative warm-up phase CD ALT (black line with squares) are plotted over the reference velocity points v_j. However, the chosen test procedures, which are based on the requirements according to GTR No. 15, are modified. Thus, to avoid measurement result scatter, the braking phase at the beginning of the warm-up phase is omitted. Instead, it is ensured that the measurement results will not contain residual brake forces. Furthermore, the minimum time requirements for the warm-up phase of 1200 s are chosen, initially.

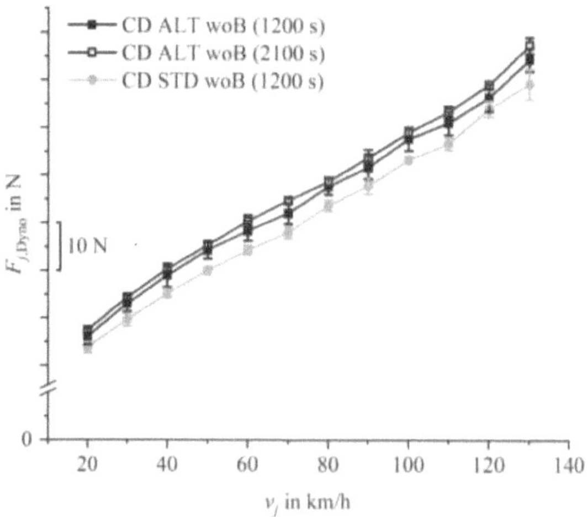

Figure 4.4: Road load determined using the test procedures with the measurement phase by deceleration and a standard warm-up phase with a duration of 1200 s CD STD woB (grey line with dots), with the alternative warm-up phase with a duration of 1200 s CD ALT woB (black line with filled squares) and with the alternative warm-up phase but with an increased warm-up duration of 2100 s CD ALT woB (black line with open squares).

It can be seen that the results for the test procedure with the standard warm-up phase (CD STD) are on average about 3.8 N lower than the results with the alternative warm-up phase (CD ALT). The main reason for the lower road loads are the higher oil temperatures of the rear wheel drive and the manual transmission, as can be seen in Figure 4.5. In this figure the temperatures of all four tyre treads and the oil temperature of the rear wheel drive, as well of the manual transmission, are plotted as bar charts.

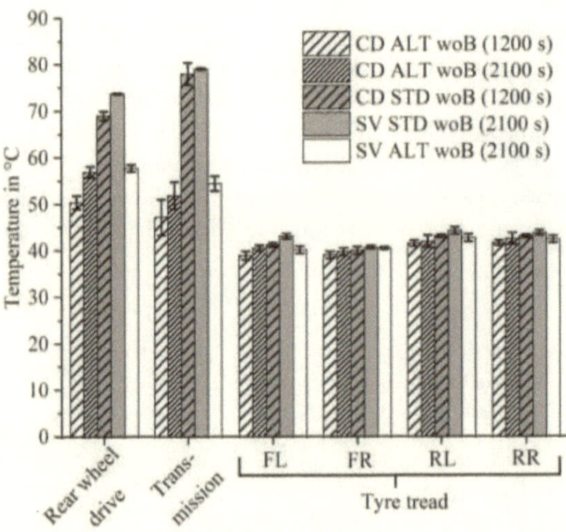

Figure 4.5: Temperatures of the tyre treads and of the oil in the rear wheel drive and in the manual transmission at the end of the warm-up phase.

This illustration clarifies that the differences in the test procedures do not have a significant effect on the temperature of the tyre treads. However, if the vehicle is driven by its own engine (standard warm-up phase), the oil temperature of the manual transmission is nearly 70 °C and of the rear wheel drive about 80 °C. This results in a temperature difference of nearly 20 °C for the rear wheel drive and over 30 °C compared to the test procedure with the alternative warm-up phase (CD ALT woB), but with the same warm-up phase duration. And as already mentioned in subsection 2.2.3, the oil viscosity decreases with increasing oil temperature [50] and thus the drivetrain losses decrease.
In the next step, the duration of the warm-up phase of the test procedure with the alternative warm-up phase is increased to reach the transmission temperatures of the standard warm-up phase. The warm-up duration was increased by 75 % from 1200 s to 2100 s. However, the transmission oil only increases by about 5 °C for the manual transmission and 6 °C for the rear wheel drive (see Figure 4.5). Furthermore, it can be seen that there is no significant difference between the road load determined with the minimum warm-up duration of 1200 s (black line with squares) and with the increased warm-up duration of 2100 s (black line with open squares). On average, the difference between these two configurations is 1.8 N, considering that the four load cells of the flat belt dynamometer have in total an accuracy of ±1.2 N [12].
Afterwards, the influence of the different measurement phases (deceleration and stabilized velocity) is investigated and illustrated in Figure 4.6. In a first step, the measurement phase with stabilized velocity and the alternative warm-up phase with a duration of 2100 s SV ALT woB (black line with triangles) is compared to the test procedure where the config-

urations for the warm-up phase are the same, but the measurement phase by deceleration CD ALT woB (black line with open squares) is chosen. The curve of the determined road load using CD ALT woB (2100 s) is the same curve as already illustrated in Figure 4.4.

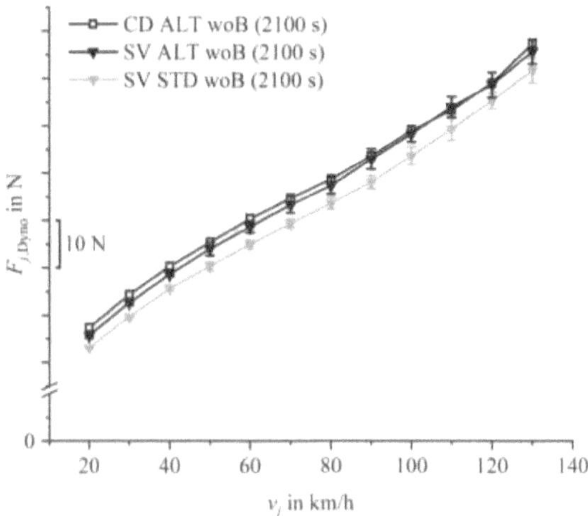

Figure 4.6: Road load determined by test procedures with the measurement phase with stabilized velocity and a standard warm-up phase with a duration of 2100 s SV STD woB (grey line with triangles), with the alternative warm-up phase with a duration of 2100 s (black line with triangles) compared to the road load determined with a measurement phase by deceleration and an alternative warm-up phase with a duration of 2100 s CD ALT woB (black line with open squares) over the reference velocity points v_j.

It can be seen that the road loads determined with SV ALT woB (black line with triangles) and with CD ALT woB (black line with open squares) are nearly at the same level. The maximum deviation is 1.7 N. This is consistent with the low differences of the tyre tread and transmission temperatures (see Figure 4.5). The maximum temperature difference can be found for the manual transmission oil and amounts to only 2.6 °C. Furthermore, it can be seen that if the alternative warm-up phase is replaced by the standard warm-up phase, the transmission oil temperatures for the test procedure SV STD oB (grey bars) are again nearly at the same level as for the test procedure CD STD woB (grey bars with stripes) (see Figure 4.5). The average road load difference between these two road loads determined with the two different warm-up phases (ALT and STD), plotted in Figure 4.6, is 3.8 N. This is similar to the results which have been obtained for the test procedures with the measurement phase by deceleration (compare Figure 4.4).

Finally, it can be concluded that the chosen measurement phase has no significant influence on the measured road load. However, there is a difference in the road load depending on the chosen warm-up phase. If the standard warm-up phase is chosen, the determined road load has lower absolute values due to the higher transmission oil temperatures (see Figures 4.5). This is also in accordance with investigations to a smaller extent made in [8].

4.1.3 Separation of rolling resistance and drivetrain losses at the flat belt dynamometer

To investigate the influencing factors on the rolling resistance and on the drivetrain losses separately and in more detail, the road load measured at the flat belt dynamometer $F_{j,\mathrm{Dyno}}$ should be separated into its two components, rolling resistance $F_{j,\mathrm{Roll}}$ and drivetrain losses $F_{j,\mathrm{Drive}}$, using the torque measurement system, as introduced in subsection 3.3.4 and developed by Untermaierhofer, Petz and Vogeler [72].

In the first step, it is investigated whether there are differences in the road load measured with and without mounted TOM (TOrque Meter, see subsection 3.3.4), as illustrated in Figure 4.7. The modified test procedure without braking phase SV ALT woB (see Figure A.1) is used. The duration of the warm-up phase is 2100 s for all measurements [73].

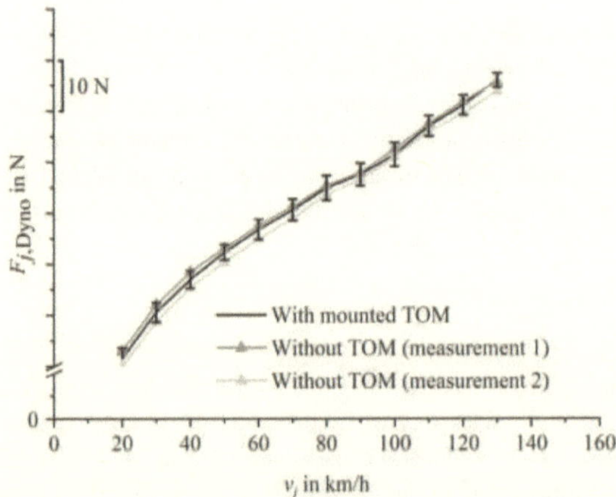

Figure 4.7: Measured road load at the flat belt dynamometer with mounted (black line) and without TOM (light/dark grey line with triangles) (adapted by permission from Isabell Vogeler and Springer Nature Customer Service Centre GmbH: Springer Nature, ATZ - Automobiltechnische Zeitschrift [73], 2020).

This figure shows the averaged flat belt dynamometer measurements (black line) over the reference velocity points v_j, where the additional measurement system TOM is mounted on the vehicle. The corresponding standard deviation of these measurements is about 1.9 N. Additionally, two further road load curves are plotted, which are measured without mounted TOM (light/dark grey line with triangles). It can be seen that the results of these measurements are in the range of the error bars of the measurements with mounted TOM. The maximum deviation between the measurements with mounted TOM and without TOM is 2.2 N and occurs at the $130\,\mathrm{km/h}$ reference velocity point. From these results, it can be inferred that the additional mounted measurement system TOM has no influence on the measured road load of a vehicle [73].

Afterwards, the absolute values of the drivetrain losses $F_{j,\mathrm{Drive}}$ measured with TOM are verified with a calibrated test bench, which is usually used to determine drivetrain losses of a vehicle. The test bench consists of two electric engines, which drive the investigated vehicle axle (front or rear) over its wheel hubs. In addition, shaft torque meters are located between the wheel hubs and the electric engines, which enables the measurement of the drivetrain losses. For the verification of TOM, the test setup showed in Figure 4.8 is used. In this case, the rear axle of a vehicle with front wheel drive (white blocks) is investigated. Additionally, the TOM (dark grey blocks) is installed between the shaft torque meters of the test bench (light grey blocks) and the wheel hubs of the vehicle (white blocks) [73].

For the verification, the test procedure illustrated in A.1, which is usually used for the determination of drivetrain losses at the test bench, is modified. In this case, after a warm-up duration of $1200\,\mathrm{s}$ at $130\,\mathrm{km/h}$, three measurement phases from the highest reference velocity of $140\,\mathrm{km/h}$ to the lowest reference velocity of $10\,\mathrm{km/h}$ in incremental steps of $10\,\mathrm{km/h}$ are conducted. This test procedure is repeated two times [73].

The averaged drivetrain losses for the rear axle of the vehicle $F_{j,\mathrm{Drive,RA}}$ as measured with the shaft torque meter of the test bench (black line) and the TOM (grey line) plotted over the reference velocity points v_j are shown in Figure 4.9 for the front axle (a) and for the rear axle (b). In addition, the difference between these two measurement systems $|\Delta F|$ is plotted using the right axis.

It can be seen that the standard deviation of the results measured with the shaft torque meters for the front axle (a) is on average about 0.2 N while it is about 0.6 N when measured with TOM. Furthermore, the averaged difference $|\Delta F|$ is 1.2 N. As already stated in [73], the averaged standard deviation of the results measured with the test bench of the drivetrain losses at the rear axle (b) is 0.3 N, and measured with TOM it is about 0.2 N. The maximum difference $|\Delta F|$ is about 0.5 N in the velocity range from $120\,\mathrm{km/h}$ to $140\,\mathrm{km/h}$. In the lower velocity range the difference is even lower [73]. In total, the differences between TOM and the test bench for drivetrain losses for both axles is relatively low when compared to the absolute value of drivetrain losses of about 50 N (see the assumption in chapter 2.2.3 and [26]). Thereby, the absolute values for the drivetrain losses measured with TOM are verified for both axles.

Concluding the results shown in Figures 4.8 and 4.9, it is possible to determine the drivetrain losses of a vehicle and to separate the result measured with the flat belt dynamometer into its two components, rolling resistance $F_{j,\mathrm{Roll}}$ and drivetrain losses $F_{j,\mathrm{Drive}}$, according

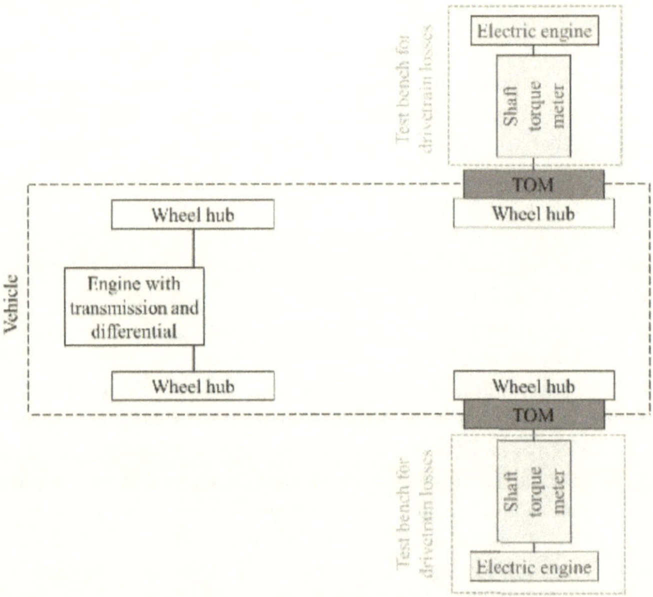

Figure 4.8: Schematic illustration of the test setup to verify TOM (dark grey blocks) with a test bench for the determination of drivetrain losses (light grey blocks) in this case for the rear axle of a vehicle with front wheel drive (white blocks) (adapted by permission from Isabell Vogeler and Springer Nature Customer Service Centre GmbH: Springer Nature, ATZ - Automobiltechnische Zeitschrift [73], 2020).

to Equation 3.11 [73]. Finally, in Figure 4.10 the resulting components, drivetrain losses $F_{j,\text{Drive}}$ (lines with filled symbols) and rolling resistance $F_{j,\text{Roll}}$ (lines with open symbols) referenced to the force measured at the flat belt dynamometer $F_{j,\text{Dyno}}$ for the front (FA, grey lines with dots) and the rear axle (RA, black lines with squares) of the vehicle are plotted over the reference velocity points v_j. It shows that in this case the rolling resistance component is about 88 % at the rear axle and for a wide velocity range about 73 % at the front axle related to the flat belt result $F_{j,\text{Dyno,RA}}$ or $F_{j,\text{Dyno,FA}}$ over the investigated velocity range. The ratio of the rolling resistance at the front axle is smaller when compared to the ratio at the rear axle, because the vehicle has front wheel drive.

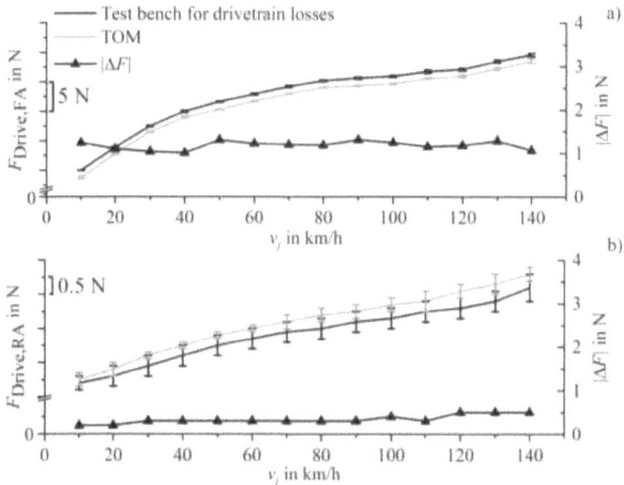

Figure 4.9: Verification of the newly developed TOM (grey line) with the test bench for drivetrain losses (black line) and the difference $|\Delta F|$ between these two measurement systems (black line with triangles) for the front axle (a) and for the rear axle (b) (extended by the front axle and adapted by permission from Isabell Vogeler and Springer Nature Customer Service Centre GmbH: Springer Nature, ATZ - Automobiltechnische Zeitschrift [73], 2020).

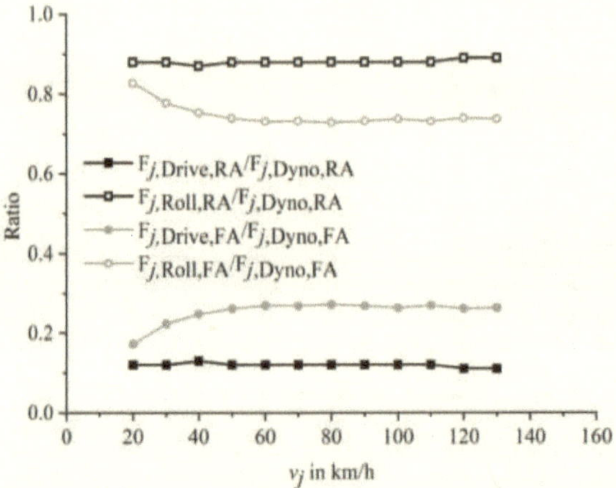

Figure 4.10: Ratios between the components rolling resistance (open symbols) or drivetrain losses (filled symbols) and the road load measured at the flat belt dynamometer for the front axle (grey lines with dots) and for the rear axles (black lines with squares).

4.1.4 Variation of tyre inflation pressure

In the next step, the influence of the tyre inflation pressure on the rolling resistance is investigated and compared with the assumption given in [26, 30] (see also Equation 2.24). This estimation states a power function between the rolling resistance and the tyre inflation pressure with an exponent of -0.4.

For this investigation the tyre inflation pressure is decreased and increased as compared to the standard tyre inflation pressure $p_{\text{Tyre,Std}}$ given by the manufacturer specifications. The road load is measured at the flat belt dynamometer using the test procedure SV ALT woB (compare Figure A.1). Furthermore, the drivetrain losses are determined using TOM (see subsection 3.3.4) and are afterwards subtracted from the road load, which is measured using the flat belt dynamometer, to obtain only the rolling resistance $F_{j,\text{Roll}}$.

In Figure 4.11 the rolling resistance $F_{j,\text{Roll}}$ dependence on the tyre inflation pressure for the front axle (solid lines) and the rear axle (dashed lines) are illustrated. For the sake of clarity, only the pressure curves for the reference velocity points v_j at $20\,\text{km/h}$ (black lines), $80\,\text{km/h}$ (dark grey lines) and $130\,\text{km/h}$ (light grey lines) are shown. The test vehicle F33 430i is used. The standard inflation pressure $p_{\text{Tyre,Std}}$ according to the manufacturer specification is 2.6 bar at the front axle and 2.9 bar at the rear axle.

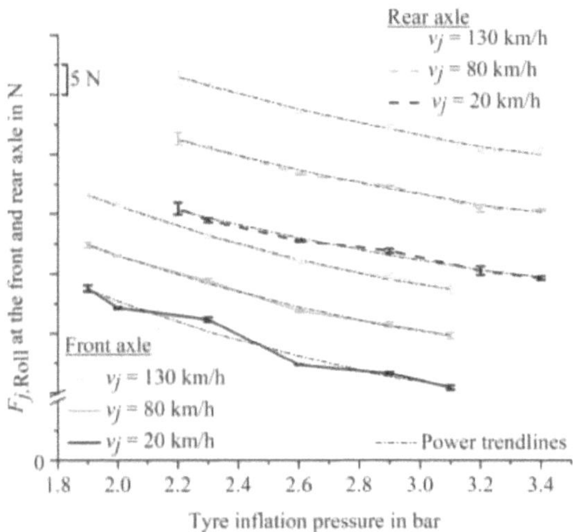

Figure 4.11: Rolling resistance $F_{j,\text{Roll}}$ depending on the tyre inflation pressure for the front axle (solid lines) and the rear axle (dashed lines) at the reference velocity points at $20\,\text{km/h}$ (black lines), $80\,\text{km/h}$ (dark grey lines) and $130\,\text{km/h}$ (light grey lines) as well as the corresponding power trendlines (dotted lines).

It can be seen that in all cases the rolling resistance $F_{j,\mathrm{Roll}}$ decreases with increasing tyre inflation pressure and with decreasing velocity, which is in accordance with the existing theoretical knowledge (see subsection 2.2.2). In the following, these results are compared with the assumption given in [26, 30] and in Equation 2.24, which describes the influence of the tyre inflation pressure on the resulting rolling resistance. According to this assumption, the influence of the tyre inflation pressure can be described with a power function and an exponent value e of -0.4. In the next step, the exponents e_{v_j} for the rolling resistances measured at the flat belt dynamometer (see Figure 4.11) are determined, depending on the reference velocity point v_j. For this determination, the the curves illustrated in Figure 4.11 are described by a power function (see the power trendlines in Figure 4.11). As the wheel load is kept constant for these investigations, Equation 2.24 results in Equation 4.1, whereby the rolling resistance $F_{j,\mathrm{Roll},p_{\mathrm{Tyre,Std}}}$ determined with the standard tyre inflation pressure $p_{\mathrm{Tyre,Std}}$ defined by manufacturer specifications is inserted for $F_{\mathrm{Roll,ISO}}$:

$$F_{j,\mathrm{Roll},p_{\mathrm{Tyre}}} = F_{j,\mathrm{Roll},p_{\mathrm{Tyre,Std}}} \cdot \left(\frac{p_{\mathrm{Tyre}}}{p_{\mathrm{Tyre,Std}}} \right)^{e_{v_j}} \qquad (4.1)$$

where:

e_{v_j}	is an exponent value describing the influence of the tyre inflation pressure on the rolling resistance $F_{j,\mathrm{Roll}}$ at the reference velocity point v_j and the investigated tyre inflation pressure p_{Tyre};
$F_{j,\mathrm{Roll},p_{\mathrm{Tyre}}}$	is the rolling resistance of the tyre with the inflation pressure p_{Tyre} in N;
$F_{j,\mathrm{Roll},p_{\mathrm{Tyre,Std}}}$	is the rolling resistance of the tyre with the standard inflation pressure $p_{\mathrm{Tyre,Std}}$ according to manufacturer specification in N;
p_{Tyre}	is the tyre inflation pressure in bar;
$p_{\mathrm{Tyre,Std}}$	is the standard tyre inflation pressure according to manufacturer specification in bar.

The exponents e_{v_j} of these power functions and the corresponding coefficient of determination r^2 are listed in Table 4.1 for the front axle and for the rear axle of the test vehicle F33 430i.

Analyzing these results, it can be concluded that the average value for all exponents e_{v_j} at the front axle and at the rear axle given in Table 4.1 is -0.42, with a standard deviation of ± 0.08. To investigate the influence of the uncertainty of the determined exponent due to its standard deviation, the resulting difference in the calculated rolling resistance, depending on the exponent, is illustrated in Figure 4.12. Therefore, the tyre inflation pressure is decreased from the standard inflation pressure $p_{\mathrm{Tyre,Std}} = 2.6\,\mathrm{bar}$ by $0.6\,\mathrm{bar}$ to $2.0\,\mathrm{bar}$ at the front axle of the test vehicle F33 430i.

Table 4.1: Exponents e_{v_j} at each reference velocity point v_j for the front axle with $p_{\text{Tyre,Std}} = 2.6\,\text{bar}$ and for the rear axle with $p_{\text{Tyre,Std}} = 2.9\,\text{bar}$ of the test vehicle F33 430i and the corresponding coefficients of the determination r^2

v_j in $\mathrm{km/h}$	Front axle		Rear axle	
	e_{v_j}	r^2	e_{v_j}	r^2
20	-0.58	0.9663	-0.35	0.9908
30	-0.55	0.9758	-0.34	0.9933
40	-0.53	0.9804	-0.34	0.9921
50	-0.51	0.9872	-0.33	0.9893
60	-0.49	0.9934	-0.33	0.9912
70	-0.50	0.9907	-0.34	0.9873
80	-0.48	0.9955	-0.34	0.9898
90	-0.47	0.9957	-0.34	0.9921
100	-0.46	0.9967	-0.33	0.9915
110	-0.45	0.9966	-0.33	0.9920
120	-0.44	0.9969	-0.39	0.9645
130	-0.44	0.9970	-0.32	0.9853

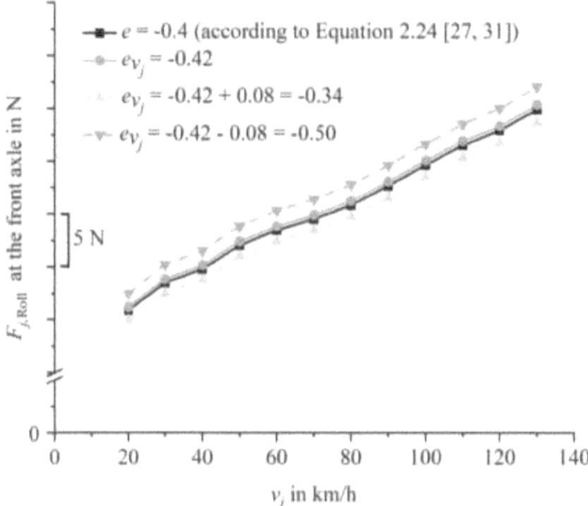

Figure 4.12: Rolling resistance $F_{j,\text{Roll}}$ depending on the used exponent value decribing the influence of the tyre pressure on the rolling resistance for the front axle of the test vehicle F30 430i.

It is pointed out that the difference in the resulting rolling resistance using the exponent value of -0.4 (black line with squares) assumed in [26, 30] and the determined exponent with a value of -0.42 (grey line with dots) is 0.4 N on average. Considering that the accuracy of the load cells of the flat belt dynamometer is at one axle in sum 0.6 N[12], this deviation is negligible. If the standard deviation of the determined exponents of ±0.08 (see Table 4.1) is additionally considered to the exponent of -0.42 (dashed lines with triangles), there is a difference of the calculated rolling resistance of about ±1.5 N on average as compared to the rolling resistance calculated with an exponent of -0.42.

Due to the high deviation of the computed exponents of the test vehicle F33 430i, the exponent is verified with two further test vehicles. The inflation pressure of the test vehicle G11 725dA with a standard inflation pressure of 2.2 bar at both axles is varied by +0.6 bar, +0.3 bar, -0.2 bar, -0.4 bar and -0.6 bar. For the third test vehicle F46 216i with a standard inflation pressure of 2.2 bar at both axles the inflation pressure is varied by about ±0.6 bar.

The computed averaged exponents e_{v_j} with their standard deviations for these three vehicles are given in Table 4.2. The single exponent values for the G11 725dA are given in Table B.1 and for the F46 216i in Table B.2 in the appendix.

Table 4.2: Computed averaged exponents e_{v_j} with their standard deviation for the test vehicles F33 430i, G11 725dA and F46 216i

Vehicle	e_{v_j}
F33 430i	-0.42 ±0.08
G11 725dA	-0.40 ±0.08
F46 216i	-0.42 ±0.04

It can be seen that for all three vehicles the exponents have on average the same magnitude as the exponent with a value of -0.4 given in [26, 30]. However, similiar to the first vehicle (F33 430i), the exponents for the other two vehicles deviate with a standard deviation between 0.04 and 0.08. Moreover, it has to be considered that the determination of the exponent e given in [26, 30] is based on rolling resistance measurements, which are executed according to ISO 8767 [75] or ISO 18164 [76]. In contrast to the measurement method on the flat belt dynamometer, there are several differences to the measurement method using the ISO standardization. The differences are listed in Table 4.3.

Table 4.3: Differences between the measurement method at the flat belt dynamometer compared to the measurement method using the ISO standardization [75, 76] for rolling resistance determination

Characteristic	Flat belt dynamometer	ISO standardization
Rolling surface	flat belt	drum with a diameter between 1.5 and 3 m
Surface texture	coated with Safety Walk^{TM}	smooth or textured with a roughness of 180 µm
Warm-up velocity	144 km/h	80 km/h
Measurement velocity	20 km/h - 130 km/h	80 km/h

Nevertheless, due to the similar exponent values these results show that the tyre inflation pressure has nearly the same influence on the rolling resistance measured at a drum test bench and at the flat belt dynamometer. Neither the rolling surface, the surface texture nor the warm-up velocity seem to have an influence on the assumption made in [26, 30]. Therefore, Equation 2.24 can be used to estimate the influence of the tyre inflation pressure on the rolling resistance at the flat belt dynamometer. However, it has to be considered that this is only an approximate estimation, since the computed exponent values listed in Table 4.2 exhibit a high standard deviation compared to the absolute value of the determined exponent e stated in [26, 30] (compare also Figure 4.12).

4.1.5 Influence of the vehicle position on the flat belts on the road load

In this subsection, the influence of the vehicle position on the flat belts on the rolling resistance and on the drivetrain losses is investigated. Besides the detailed definition of the different measurement methods, which are allowed using the wind tunnel method (see chapter 2.3.2), also the position of the vehicle at the flat belts is specified in the GTR No. 15. It is specified that the vehicle shall be aligned within a tolerance of ±0.5 degrees of the rotation around its z axis at the flat belts [10]. To ensure that this explicit specification is fulfilled for all measurements, the vehicle front is fixed at the front tow hook and is then centered at the flat belts with the aid of a laser (front laser). Afterwards, the vehicle rear is centered with the aid of a second laser (rear laser). Both lasers are orientated so that they both mark the centre line of the flat belt dynamometer (see Figure 4.13).

To investigate the influence of the vehicle position on the flat belt dynamometer on the measured road load, different rotation angles β of the test vehicle F46 216i are set up:

- β = -0.01°: Both the vehicle front and the vehicle rear are centered with the aid of the front and the rear laser.

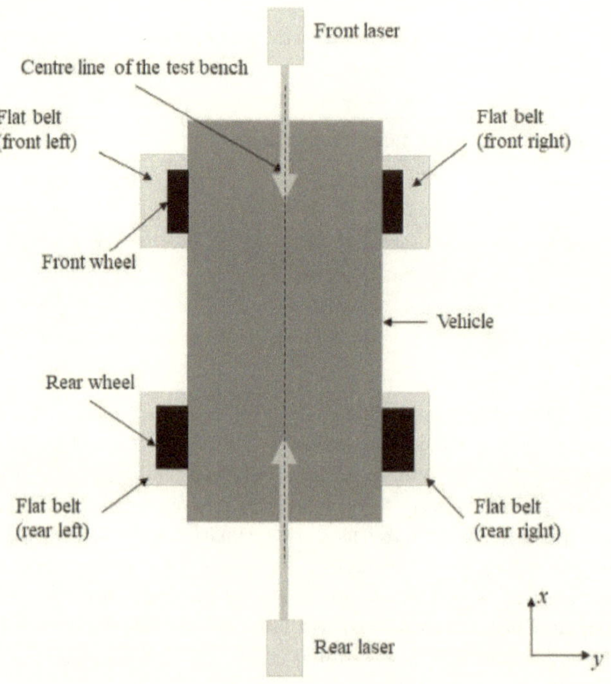

Figure 4.13: Schematic illustration of the laser orientation at the flat belt dynamometer to centre the vehicle at the flat belts.

- $\beta = +0.15°$: The vehicle front is fixed at the front tow hook and is centered at the flat belts with the aid of the front laser. Afterwards, the vehicle is driven at a low velocity (in this case at $5\,\mathrm{km/h}$) and is then fixed at the rear tow hook.

- $\beta = +0.52°$: The vehicle front is centered. Afterwards, the vehicle rear is positioned with the maximum allowed rotation angle β in positive direction according to GTR No.15.

- $\beta = -0.52°$: The vehicle front is centered. Afterwards, the vehicle rear is positioned with the maximum allowed rotation angle β in negative direction according to GTR No.15.

For the calculation of the rotation angle β, the angle is initially shifted parallel to the outer edge of the front wheel, as can be seen in Figure 4.14. Consequently, the rotation angle β is defined by the arc sine of the distance difference of the front and of the rear axle $\Delta d_{\mathrm{FA,RA}}$ and the wheel base of the vehicle l_{WB}. The difference of the front and rear axles $\Delta d_{\mathrm{FA,RA}}$ is the delta between the outer edge of the front wheel d_{FA} and of the rear wheel d_{RA} related to a reference edge and the delta value Δd. The delta value Δd is determined by the different tyre widths and tracks at the front and rear axles. The corresponding equations are defined as:

$$\beta = \arcsin\left(\frac{\Delta d_{\mathrm{FA,RA}}}{l_{\mathrm{WB}}}\right) \cdot \frac{180}{\pi} \tag{4.2}$$

with:

$$\Delta d_{\mathrm{FA,RA}} = d_{\mathrm{FA}} - d_{\mathrm{RA}} - \Delta d \tag{4.3}$$

and

$$\Delta d = w_{RT} - w_{FT} + {}^{1}\!/\!{}_{2}\left(l_{\mathrm{Track,RA}} - l_{\mathrm{Track,FA}}\right) \tag{4.4}$$

where:

β	is the rotation angle of the vehicle around its z axis in degrees;
$\Delta d_{\mathrm{FA,RA}}$	is the difference of the front and rear axle in mm;
l_{WB}	is the wheel base of the vehicle in mm;
d_{FA}	is the distance of the outer edge of the front wheel to the reference edge in mm;
d_{RA}	is the distance of the outer edge of the rear wheel to the reference edge in mm;
Δd	is a delta value due to different tyre widths and tracks at the front and rear axles in mm;
w_{RT}	is the tyre width of the rear tyre in mm;
w_{FT}	is the tyre width of the front tyre in mm;
$l_{Track,FA}$	is the track at the front axle in mm;
$l_{Track,RA}$	is the track at the rear axle in mm.

In Figure 4.15, the drivetrain losses $F_{j,\mathrm{Drive}}$ at the front (FA, open symbols) and at the rear axle (RA, filled symbols) measured with TOM (see subsection 3.3.4) for different rotation angles β of -0.52° (grey lines with diamonds), -0.01° (black lines with dots), +0.15° (dark grey lines with squares) and +0.52° (light grey lines with triangles) are plotted with the corresponding standard deviations over the reference velocity points v_j. For both the front and the rear axle, there is no clear correlation between the rotation angle and the drivetrain losses. The results for the different rotation angles deviate about $\pm 1.5\,\mathrm{N}$ for each axle. This result is plausible, since due to the small magnitude of rotation angles, the cooling effect of the fan on the drivetrain is not expected to be significant.

In Figure 4.16 the resulting rolling resistance $F_{j,\mathrm{Roll}}$ at the front (a) and at the rear axle (b) determined using Equation 3.11 for the different rotation angles β of -0.52°(grey lines with diamonds), -0.01° (black lines with dots), +0.15° (dark grey lines with squares) and +0.52° (light grey lines with triangles) are plotted with the corresponding standard deviations over the reference velocity points v_j.

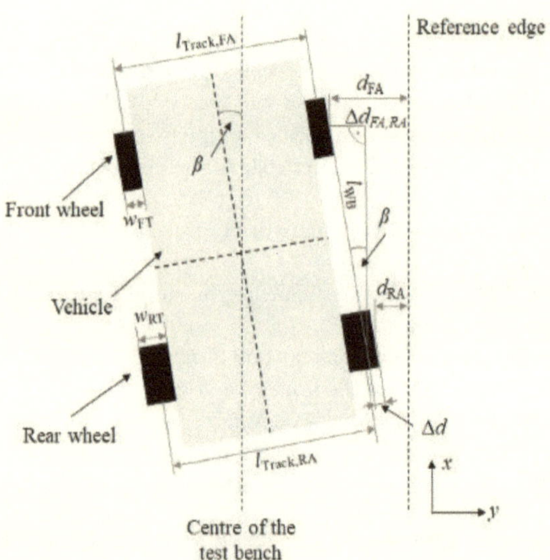

Figure 4.14: Schematic illustration of the vehicle positioned with a rotation angle β at the test bench, whereby the misalignment is exaggerated.

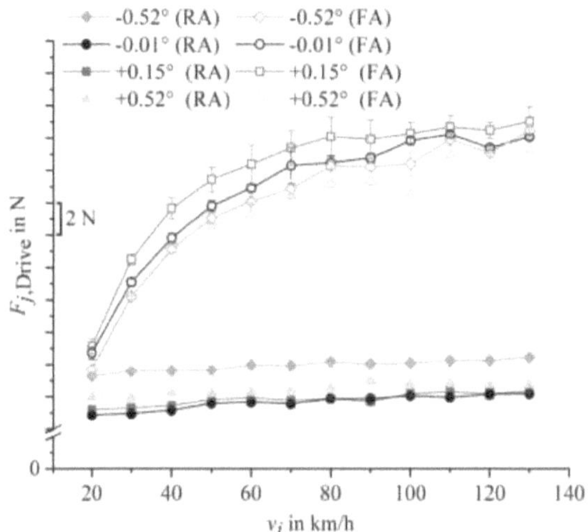

Figure 4.15: Influence of the rotation angle β of -0.52°(grey lines with diamonds), -0.01° (black lines with dots), +0.15° (dark grey lines with squares) and +0.52° (light grey lines with triangles) on the drivetrain losses $F_{j,\mathrm{Drive}}$ of the front (FA, open symbols) and rear axle (RA, filled symbols) plotted over the reference velocity points v_j.

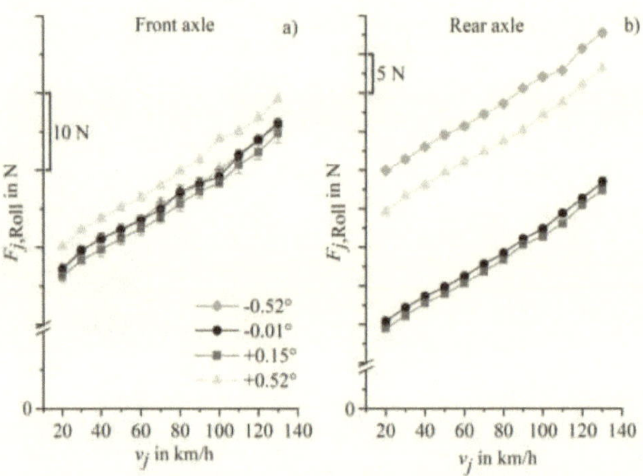

Figure 4.16: Influence of the rotation angle β of -0.52°(grey line with rhombus), -0.01° (black line with dot), +0.15° (dark grey line with square) and +0.52° (light grey line with triangle) on the rolling resistance $F_{j,\mathrm{Roll}}$ of the front (a) and of the rear axle (b) plotted over the reference velocity points v_j.

As the front of the vehicle is always centered with a laser, there is again no clear tendency of the rotating angle influencing the rolling resistance at the front axle (see Figure 4.16 a). The road load determined with the rotation angles of -0.52°, -0.01° and +0.15° differ with an average standard deviation of 0.8 N, whereby the maximum deviation with a value of 1.2 N occurs at the velocity of 100 km/h. Only the road load measured at a rotation angle of +0.52° is about 3 N higher than the other road loads. Furthermore, the difference between the rotation angle of -0.01° and +0.15° is only about 1 N at the rear axle. Considering the accuracy of the flat belt dynamometer load cell for one axle of about 0.6 N[12], this influence is negligible. In contrast, the rolling resistance at the rear axle (see Figure 4.16 b) is increased by about 15 N for a rotation angle of +0.52°, and even 20 N for -0.52°. This increase can be attributed to the fact that because of the rotation angle the tyre position at the rear is inclined with respect to the driving direction, which results in an additional toe of 0.5° at each rear tyre. Therefore, a theoretical estimation is made in the following, to verify the increased rolling resistance measured at the flat belt dynamometer. For a wheel load of about 3747 N at the rear axle of the F46 216i a minimum cornering stiffness of 1452 N/° and a maximum cornering stiffness of 1840 N/° can be assumed, using the information illustrated in Figure 2.16 and given in [31]. The resistance due to the additional toe $F_{\mathrm{Roll,Toe}}$ can then be estimated with Equation 2.20 and lies in this case between 25 N and 32 N. However, there is a difference between theoretical calculated force values and the measured forces, but the values are of the same magnitude. For a better estimation, the real stiffness values are necessary, which were not available for this tyre type. Furthermore, there is no symmetrical behaviour for the rolling resistance at the rear axle especially in the case the vehicle is inclined with a rotation angle of -0.52° and +0.52°. For this investigation, no new tyres are used. Instead they are the same tyres which are also used for the road load determination on a proving ground (coastdown method, see subsection 2.3.1). As it can be seen in Figure 3.1, the vehicle has to drive a left turn between each measurement phase. Therefore, an asymetrical tyre abrasion can be assumed, which may result in asymmetrical results in the case the vehicle is inclined by an additional rotation angle at the test bench.

4.1.6 Road load determination at different ambient temperatures

To investigate the influence of different ambient temperatures T_{Amb} on the resulting road load of the test vehicle F46 216i, the standard ambient temperature T_{Amb} of 23 °C at the flat belt dynamometer is reduced to 15 °C and to 10 °C. Prior to each measurement at the different ambient temperatures the inflation pressure of each tyre is set to 2.2 bar. With the aid of TOM (see subsection 3.3.4), it is additionally possible to assign the effect on the road load to the rolling resistance $F_{j,\mathrm{Roll}}$ and the drivetrain losses $F_{j,\mathrm{Drive}}$.

In Figure 4.17 the resulting temperatures of the tyre treads for all four tyres (FL = front left, FR = front right, RL = rear left and RR = rear right) and of the transmission oil are illustrated for the different ambient temperatures 23 °C (black bars), 15 °C (grey bars) and 10 °C (light grey bars) at the end of the warm-up phase. The tyre tread temperatures are measured with the measurement technology described in subsection 3.3.1.

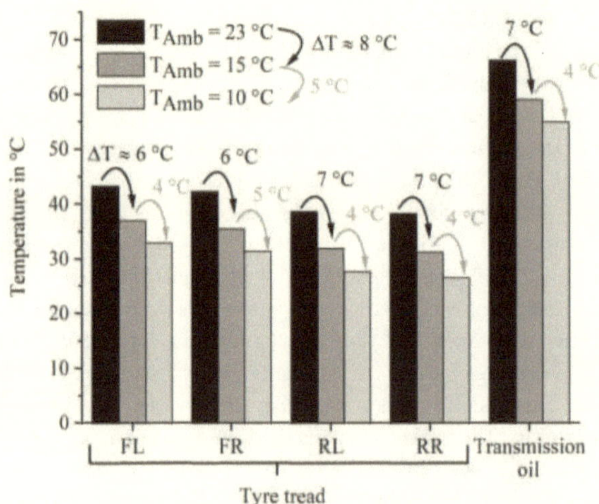

Figure 4.17: Temperatures of the tyre treads for all four tyres (FL = front left, FR = front right, RL = rear left and RR = rear right) and of the transmission oil for the different ambient temperatures 23 °C (black bars), 15 °C (grey bars) and 10 °C (light grey bars) at the end of the warm-up phase.

It can be seen that with decreasing ambient temperature, both the tyre tread and the transmission oil temperatures decrease by nearly the same amount. In Figure 4.18 the percentage deviations of the rolling resistance $F_{j,\text{Roll}}$ at 15 °C (grey line with dots) and at 10 °C (light grey line with triangles), normalized to the rolling resistance determined at an ambient temperature of 23 °C, are plotted over the reference velocity points v_j. In this case, the flat belt measurement results and therefore also the rolling resistances are not corrected to the reference temperature of 20 °C, as described in Equation 2.53.

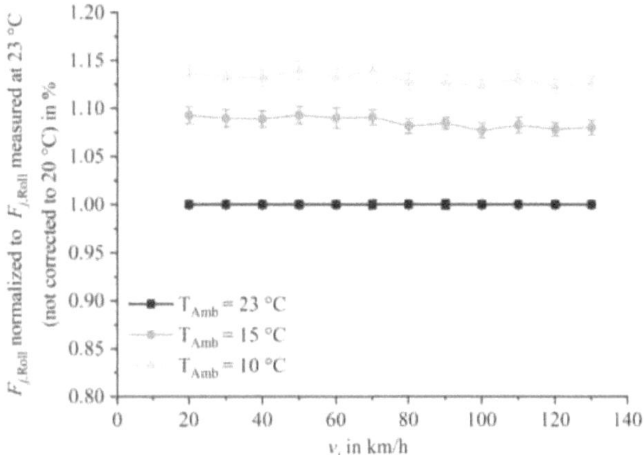

Figure 4.18: Influence of the ambient temperature at the flat belt dynamometer on the uncorrected rolling resistance ($T_\mathrm{Amb} = 23\,°\mathrm{C}$: black line with squares; $15\,°\mathrm{C}$: grey line with dots; $10\,°\mathrm{C}$: light grey line with triangles) normalized to the rolling resistance measured at $23\,°\mathrm{C}$.

It can be seen that with decreasing ambient temperature of about $8\,°\mathrm{C}$ from $23\,°\mathrm{C}$ to $15\,°\mathrm{C}$, the rolling resistance is increased by about $9\,\%$. A further decrease to an ambient temperature of $10\,°\mathrm{C}$ results in a reduction of the tyre tread temperatures by about $10\,°\mathrm{C}$ to $11\,°\mathrm{C}$ and an increase of rolling resistance of about $13\,\%$ as compared to the measurements made at $23\,°\mathrm{C}$. The standard deviations for each measurement point are smaller than $\pm1\,\%$, except for the measurement point at $70\,\mathrm{km/h}$ and $10\,°\mathrm{C}$. The standard deviation has at this point a value of about $\pm1.2\,\%$. Therefore, the results described in [19], which show that the rolling resistance coefficient is increased by about $43\,\%$ due to a decrease of the tyre tread temperature of about $3.5\,°\mathrm{C}$, cannot be confirmed. However, it should be noted that the measurements in [19] are executed on a public road and not at a test bench under laboratory conditions. In contrast, the assumption made in [26] that a variation of the ambient temperature of $1\,°\mathrm{C}$ corresponds to a variation in the rolling resistance of $0.6\,\%$ in the temperature range of $10\,°\mathrm{C}$ to $40\,°\mathrm{C}$, results in an underprediction of the resulting rolling resistance when compared to the results presented in Figure 4.18. Thus, due to the assumption made in [26] a reduction of the ambient temperature of $8\,°\mathrm{C}$ (for example from $23\,°\mathrm{C}$ to $15\,°\mathrm{C}$) only leads to an increase of the rolling resistance of about $4.9\,\%$ and a reduction of T_Amb of $13\,°\mathrm{C}$ to an increase of about $8.1\,\%$. However, the rolling resistance measurements presented in [26] are executed at a test drum with at least a diameter of $1.7\,\mathrm{m}$ and a measurement velocity of $80\,\mathrm{km/h}$. In addition, the surface of the test drum is made of steel and is not coated with Safety Walk^{TM} as are the WDUs of the flat belt dynamometer [26, 75, 76]. Generally, at lower ambient temperatures, the tyre tread temperatures decrease which results in an increased rolling resistance. A reason for this can be found in higher tyre stiffness values and in more deformations due to a decreased internal air pressure (see subsection 2.2.2). Addtionally, in [41] it is described that due to an increased tyre core temperature of 10 °C, the cornering stiffness is decreased in between 3 % and 4 %. However, the influence of the temperature dependency of the cornering stiffness on the resulting rolling resistance is comparatively small related to the absolute value of the rolling resistance: In the following, a tyre with a toe of 20' and a cornering stiffness of 1840 ^{N}/° at a tyre tread temperature of 40°C is assumed. Due to the decrease of the tyre tread temperature by about 10°C it is assumed that the cornering stiffness is increased by about 4.2% (assuming the maximum value of dependency given in [41]) to 1917 ^{N}/°. Using Equation 2.20, this only results in an increased rolling resistance of about 0.15 N for one tyre.[18]

Moreover, not only the tyre tread temperatures are influenced by the change in ambient temperature, but also the oil in the transmission gears, as can be seen in Figure 4.17. The influence of the ambient temperature on the drivetrain losses at the front and at the rear axle over the velocity is plotted in Figure 4.19. The values are again normalized to the drivetrain losses determined at 23 °C for the front and for the rear axle (black line with squares), respectively.

Since the test vehicle has a front wheel drive, at the front axle a reduction of the ambient temperature from 23 °C to 15 °C results in an increase of the drivetrain losses of up to 24 % at 20 ^{km}/h and 9 % at 130 ^{km}/h. A further decrease in ambient temperature of up to 10 °C leads to 34 % higher drivetrain losses at a velocity of 20 ^{km}/h and 18 % at 130 ^{km}/h. In contrast, the influence of the ambient temperature on drivetrain losses of the rear axle (15 °C: grey dashed line with dots; 10 °C: light grey dashed line with dots) is low and exhibits no clear tendency. On average, the drivetrain losses increase by about 14 % in the investigated velocity range, if the ambient temperature is decreased from 23 °C to 10 °C. Although, the percentage proportion of the influence of the ambient temperature on the drivetrain losses is higher than for the rolling resistance (compare Figure 4.18), it has to be considered that the ratio between drivetrain losses and rolling resistance is only about 26 % on average for this vehicle (see in Figure B.1 in the appendix). Therefore, the influence of the ambient temperature on the absolute values of the drivetrain losses is lower than for the rolling resistance.

Until this point, the results measured with the flat belt dynamometer are not corrected to the reference ambient temperature of 20 °C. To investigate the effect of the rolling resistance factor K_0, which is applied to correct to reference ambient temperature conditions (see in subsection 2.3.2 and Equation 2.53), the averaged corrected road load measured with the flat belt dynamometer at the different ambient temperatures 23 °C (black line with

---

[18]According to [41], it is assumed that the tyre core temperature is equal to the inner air temperature. Furthermore, according to the findings presented in Figure 4.55 it is assumed that a decrease of the tyre air temperature is accompanied by a decrease of the tyre tread temperature of the same amount.

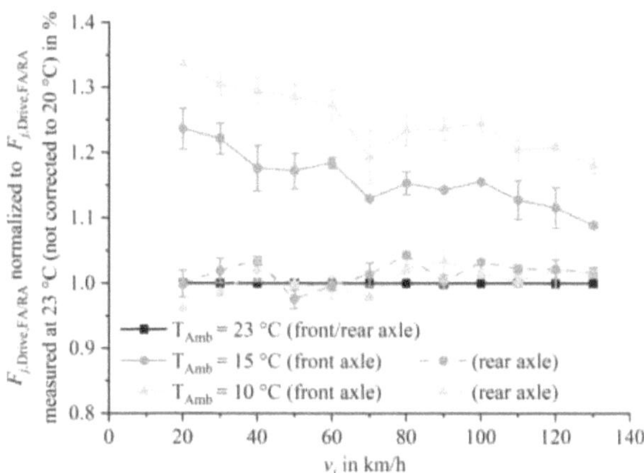

Figure 4.19: Influence of the ambient temperature at the flat belt dynamometer on the drivetrain losses ($T_{\mathrm{Amb}} = 23\,°\mathrm{C}$: black line with squares; $15\,°\mathrm{C}$: grey line with dots; $10\,°\mathrm{C}$: light grey line with triangles) normalized to the drivetrain losses measured at $23\,°\mathrm{C}$ (front axle: solid lines; rear axle: dashed lines).

squares), $15\,°\mathrm{C}$ (grey line with dots) and $10\,°\mathrm{C}$ (light grey line with triangles) are normalized to the corrected road load measured at $23\,°\mathrm{C}$ and are then compared in Figure 4.20.

It is pointed out that the measurements executed at an ambient temperature of $23\,°\mathrm{C}$ deviate about the median value with a maximum difference of $1.8\,\%$. The corrected results for the measurements made at $15\,°\mathrm{C}$ and $10\,°\mathrm{C}$ will lead to a slightly higher result for the road load at the reference velocity points of $20\,\mathrm{km/h}$ and $30\,\mathrm{km/h}$. The maximum deviation referenced to the corrected $23\,°\mathrm{C}$ measurement is $0.6\,\%$. In the higher velocity range, the $10\,°\mathrm{C}$ and $15\,°\mathrm{C}$ measurements deviate by maximum of about $1.4\,\%$ from the $23\,°\mathrm{C}$ measurement. However, it has to be emphasized that the standard deviations for all three configurations are overlapping. Therefore, it is not possible to make a clear statement about the presented deviation. Instead, it can be noted that the corrected road load measured at different ambient temperatures at the test bench will lead to similar road loads within the scope of the measurement uncertainties of the test bench.

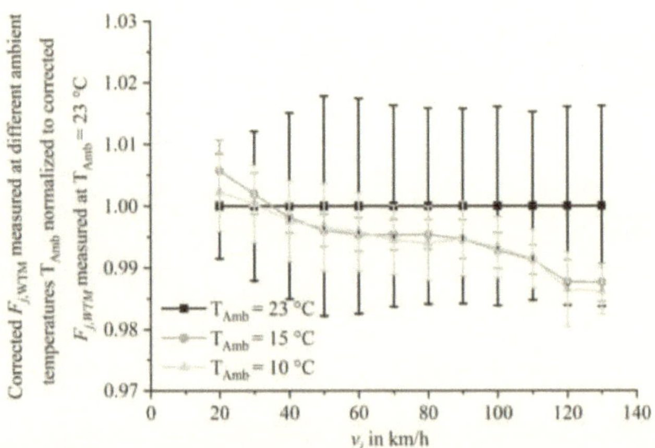

Figure 4.20: Averaged corrected road load measured at the flat belt dynamometer at the different ambient temperatures 23 °C (black line with squares), 15 °C (grey line with dots) and 10 °C (light grey line with triangles) with their corresponding standard deviations normalized to the corrected road load measured at 23 °C and plotted over the reference velocity points v_j.

4.1.7 Summary

- The airstream of the cooling fan in front of the vehicle has no influence on the force measurement at the flat belt dynamometer.

- If the standard warm-up phase is chosen, which means that the vehicle is driven by its own engine, the transmission oil temperatures are higher. This results in a lower road load when compared to the measured road load using an alternative warm-up phase (The vehicle is driven by the test bench).

- The chosen measurement phase (deceleration or stabilized velocity) has no influence on the measured road load.

- Using the newly developed torque measurement system TOM, it is possible to separate the flat belt dynamometer result into its components rolling resistance and drivetrain losses.

- Equation 2.24, which estimates the influence of the tyre inflation pressure on the resulting rolling resistance, is also valid for the rolling resistances measured with the flat belt dynamometer. However, this equation provides only an approximate estimation.

- If the vehicle stands with a rotation angle of -0.52° on the flat belt dynamometer, the rolling resistance at the rear axle is increased by about 20 N at the investigated velocity points, when compared to the rolling resistance of a vehicle which is centered at the front and at the rear on the flat belt dynamometer. The increase can be explained by the additional toe of about 0.5° at each rear tyre of the vehicle.

- The difference in rolling resistance between the rotation angle of -0.01° (vehicle front and rear centered) and +0.15° (only vehicle front centered) is on average about 2.2 N in total. Considering the accuracy of 1.2 N of the load cells at the flat belt dynamometer in total[12], this influence is low.

- The rolling resistance of the vehicle is increased by about 13 %, if the ambient temperature at the test bench is decreased by about 13 °C from 23 °C to 10 °C.

- The drivetrain losses are increased by about 14 %, if the ambient temperature at the test bench is decreased by about 13 °C from 23 °C to 10 °C.

4.2 Correlation between the coastdown method and the wind tunnel method extended

In this section, the additional correction factor K_{ext} for the newly developed wind tunnel method extended (see subsection 3.1.2) is determined using five different vehicles. Afterwards, this correction factor is verified with further eight vehicles. In the last subsection, the differences in cycle energy demand between the coastdown method and the wind tunnel method extended are investigated in detail.

4.2.1 Determination of the additional correction factor K_{ext} of the wind tunnel method extended

In subsection 2.2.2, several influencing factors on the rolling resistance are given, which are not considered with the correction terms given in GTR No. 15 (compare subsection 2.3.2). Especially the difference of the rolling resistance depending on the road/belt surface and the surface texture have an enormous influence on the absolute value of the rolling resistance. The road surface of the proving ground for the coastdown method is asphalt and concrete. In contrast, the steel belts of the flat belt dynamometer for the wind tunnel method are coated with Safety Walk^{TM}. In [38], it is found that the rolling resistance for different wheel loads on asphalt is increased by about 20 % to 30 % on average when compared to a surface with Safety Walk^{TM}. Additionally, the surface at the flat belt dynamometer is very even. On the other hand, on a real road surface unevenness and more rough surface texture can occur, which can result in up to 45 % higher rolling resistance compared to a very smooth and even surface [37]. However, in the GTR No. 15 it is specified that the road surface of the test track shall be flat, even, clean and dry [10]. Therefore, a further increase of the rolling resistance of about 45 % cannot be expected. Nevertheless, the correction terms for the wind tunnel method according to GTR No. 15 are extended in the presented study. The additional correction term and its corresponding correction factor K_{ext} are explained in subsection 3.1.2 *Wind tunnel method extended* and with Equation 3.5.

All flat belt dynamometer measurements are conducted using the test procedure according to WLTP, with braking phase, alternative warm-up phase and measurement procedure with stabilized velocity (SV ALT wB), as shown in Figure 2.28. In addition, the torque meter measurement method introduced in subsection 3.3.4 is used to separate the flat belt dynamometer result $F_{j,\mathrm{Dyno}}$ into its two components, rolling resistance $F_{j,\mathrm{Roll}}$ and drivetrain losses $F_{j,\mathrm{Drive}}$.

To determine the additional correction factor K_{ext}, the difference in cycle energy demand between the Wind Tunnel Method extended (WTMext) and the CoastDown Method (CDM) is calculated for five different vehicles using Equation 3.6, whereby the additional correction factor K_{ext} is initially set to 1. Afterwards, the value of K_{ext} is chosen in such a way that the average of the difference in cycle energy demand ($\epsilon_{\mathrm{WTMext}}$) for all chosen vehicles and for the two investigated driving cycles (WLTC and ARTEMIS European driving cycle) is nearly zero. For the determination of the optimum K_{ext}, five different vehicles are selected, which cover all types of the following categories:

- Vehicle class: compact, middle and luxury class

- Transmission type: manual (MT) and automatic transmission (AT)

- Wheel drive: front (FWD), rear (RWD) and all wheel drive (AWD)

In Table 4.4 the differences in energy demand between the wind tunnel method extended and the coastdown method are shown for the vehicles G11 725dA, F46 216i, F33 430i, F21 120iA and G15 M850iA. With the additional correction factor $K_{ext} = 1.34$, for both investigated cycles the average of the ϵ_{WTMext} values for the five vehicles is nearly zero (0.2 % for WLTC and -0.1 % for the ARTEMIS European driving cycle), which fulfills the requirements previously made. It can be seen that only the F33 430i and the F21 120iA have positive ϵ_{WTMext} values. A positive value means that the road load determined by the wind tunnel method extended is overpredicted when compared to the road load measured with the coastdown method on the road. In contrast, if a road load measured with the coastdown method is higher as compared to the road load determined with the wind tunnel method extended, the ϵ_{WTMext} value is negative. Additionally, all differences lie between ± 5 %, which is an acceptance criterion for example for the wind tunnel method according to GTR. No. 15 [10]. The greatest deviations to the coastdown measurements can be observed for the vehicles G11 725dA and F21 120iA. The reasons for this are explained in subsection 4.2.3.

Table 4.4: Differences in energy demand between the wind tunnel method extended with $K_{ext} = 1.34$ and the coastdown method for the vehicles chosen for the method development calculated on the basis of the WLTC and the ARTEMIS European driving cycle

Vehicle	Vehicle characteristics			ϵ in %	
	Vehicle class	Transmission type	Wheel drive	WLTC	ARTEMIS
G11 725dA	luxury	AT	RWD	-3.2	-3.8
F46 216i	compact	MT	FWD	-0.2	-0.2
F33 430i	middle	MT	RWD	1.2	1.2
F21 120iA	compact	AT	RWD	3.5	2.7
G15 M850iA	luxury	AT	AWD	-0.4	-0.5
Average of ϵ in %				0.2	-0.1

Note: In [60] an extended method is already presented, to reach the road load level, which exists on a real road. However, this method is only optimized for the test vehicle F46 216i and, therefore, is different to the method presented in this study.

4.2.2 Verification of the wind tunnel method extended

So far, the optimum value for K_{ext} has been determined only on the basis of five vehicles. In the following, the wind tunnel method extended with the additional correction factor $K_{ext} = 1.34$ is verified with further eight vehicles. These vehicles cover the vehicle categories as previously defined in subsection 4.2.1. The results are shown in Table 4.5.

Table 4.5: Differences in energy demand between the wind tunnel method extended with $K_{ext} = 1.34$ and the coastdown method to verify the additional correction term with $K_{ext} = 1.34$ calculated on the basis of the WLTC and the ARTEMIS European driving cycle

Vehicle	Vehicle characteristics			ϵ in %	
	Vehicle class	Trans-mission type	Wheel drive	WLTC	ARTEMIS
G20 330iA	middle	AT	RWD	-2.7	-2.4
G30 530e iPerformance	middle	AT	RWD	0.0	-0.6
G07 X7 M50dA	luxury	AT	RWD	-2.3	-2.7
J29 SPX30iA	middle	AT	RWD	-1.6	-1.2
G12 750Ld	luxury	AT	AWD	2.2	2.1
G14 M850iA	luxury	AT	AWD	1.7	1.0
F60 Countryman	compact	MT	FWD	1.9	1.6
G30 518d	middle	MT	RWD	3.2	2.6
Average of ϵ in %				0.3	0.1

It can be seen that the differences in cycle energy demand again lie between $\pm 5\%$ for all vehicles and for both driving cycles (see acceptance criterion according to GTR No. 15 [10]). Furthermore, the average value of the differences is nearly zero. They are 0.3 % for the WLTC and 0.1 % for the ARTEMIS European driving cycle. Thus, the wind tunnel method extended shows good results for the differences in cycle energy demand and is verified with further eight vehicles. Therefore, it is used in the presented form in this study for all further investigations.

However, a disadvantage of the wind tunnel method extended is that the flat belt dynamometer result, $F_{j,\mathrm{Dyno}}$, has to be divided into its components rolling resistance $F_{j,\mathrm{Roll}}$ and drivetrain losses $F_{j,\mathrm{Drive}}$. For this purpose, the torque meter TOM introduced in subsection 3.3.4 has to be mounted between the wheel hub and the rim of the investigated vehicles, which requires an increased effort for every single measurement.

Additionally, it is pointed out that for the newly developed wind tunnel method extended, only the rolling resistance is corrected. In Figure 4.21 the ratios between the drivetrain losses $F_{j,\mathrm{Drive}}$ and the road load measured at the flat belt dynamometer $F_{j,\mathrm{Dyno}}$ of the different vehicles are illustrated. It can be seen that for all vehicle types the rolling resistance part has the highest influence on the result of the flat belt dynamometer measurement.

The part of the drivetrain losses is mostly smaller than 40 % of the flat belt result $F_{j,\mathrm{Dyno}}$.

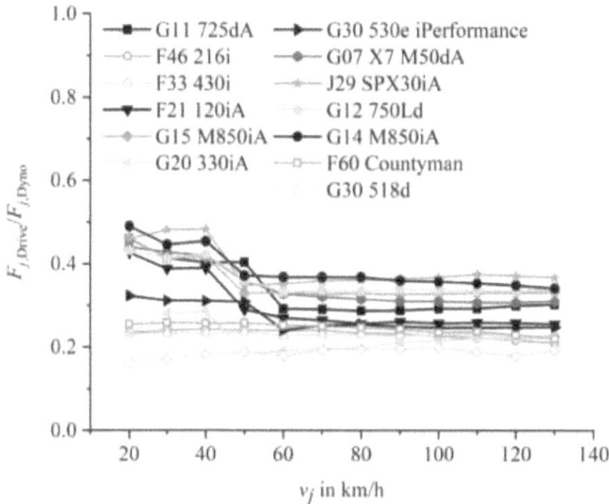

Figure 4.21: Ratios between the drivetrain losses $F_{j,\mathrm{Drive}}$ and the road load measured at the flat belt dynamometer $F_{j,\mathrm{Dyno}}$ of the vehicles over the reference velocity points.

Additionally, due to the differences in road load depending on the chosen test procedure, which are investigated in subsection 4.1.2, it is shown that the difference in road load due to different transmission oil temperatures has a relatively low value of about 3.8 N (measured with a manual transmission), considering also the assumption that the internal frictional forces of the drivetrain for an average passenger car are about 50 N [26]. Therefore, a further correction term for the drivetrain losses is omitted in the presented study.

Moreover, in Figure 4.21 the ratios for vehicles with a manual transmission are plotted with open symbols. For these vehicles a nearly linear dependency can be assumed. In contrast, for vehicles with an automatic transmission, an increase of the drivetrain losses in the velocity range from $20\,\mathrm{km/h}$ to about $50\,\mathrm{km/h}$ can be observed. All automatic transmissions of the investigated vehicles are 8HP transmissions from the company ZF Friedrichshafen AG. Although the automatic transmission is placed in neutral during the road load determination, the control device of the transmission shifts into the gears appropriate for the actual vehicle velocity[19]. In Figure 2.23, it can be seen that the efficiency of the gears generally decreases with decreasing gear, which results in increased

---

[19]Department of Mechanics and Base Functions Transmissions of the BMW Group. Influence of the gear shifting on drivetrain losses of automatic transmissions: e-mail, 20.03.2020

drivetrain losses. These losses are included in the road load, because the shift elements are located behind the clutch viewed from the engine side (see Figure 2.22) and, therefore, they explain the increase in the drivetrain losses for the 8HP automatic transmissions of ZF Friedrichshafen AG in the velocity range smaller than $60\,\mathrm{km/h}$, as can be seen in Figure 4.21.

Note: The road load determined with the coastdown method according to GTR No. 15 and the aerodynamic drag for the wind tunnel method extended for the vehicles G11 725dA sDrive, F21 120iA, G01 X3 20iA xDrive, G07 X7 M50dA, J29 SPX30iA, G12 750Ld xDrive and G14 M850iA xDrive are not measured by the author of this study. Instead, the road load values are from a BMW Group internal resource.

4.2.3 Analysis of the differences in cycle energy demand between the coastdown method and the wind tunnel method extended

In subsection 4.2.1, it is shown that the highest deviation in the difference in cycle energy demand is observed for the vehicles G11 725dA and F21 120iA. The reason can be seen in Figure 4.22. In this illustration, the ratios of the road load determined with the wind tunnel method extended with $K_{\mathrm{ext}} = 1.34$, related to the coastdown method, are plotted over the reference velocity points v_j for the five vehicles used in subsection 4.2.1.

It becomes clear that the largest differences in the road load between the coastdown method and the wind tunnel method extended occur mainly in the low velocity range for the F21 120iA (grey line with triangles) and mainly in the high velocity range for the G11 725dA (black line with squares). Furthermore, the absolute difference for the energy demand calculated with the ARTEMIS European driving cycle is 0.6 percentage points higher for the G11 725dA as compared to the difference in energy demand calculated with the WLTC. This can be explained by the different cycle velocity distribution for the WLTC and the ARTEMIS European driving cycle, as shown in Table 2.8 in subsection 2.4.2. This table clarifies that the part of velocities over $90\,\mathrm{km/h}$ amounts to 15.2 % for the WLTC and 26.3 % for the ARTEMIS European driving cycle. And it is exactly for this velocity range that the difference in road load between the wind tunnel method extended and the coastdown method for the G11 725dA is constantly about 7.5 %. Moreover, one third of the total energy demand due to the ARTEMIS European driving cycle is related to about the last $598\,\mathrm{s}$ ($\approx 12\,\mathrm{min}$) of the driving cycle. This corresponds to a share of about 19 % of the total cycle duration, which is illustrated in Figure 4.23. In this figure, the energy demands for both cycles (WLTC and ARTEMIS European driving cycle) are summed up from the start time ($t_i = 0$) until a certain time step t_i and are then related to the total energy demand of the cycle E_{WTMext} for the specific vehicle. In contrast to the ARTEMIS European driving cycle, one third of the

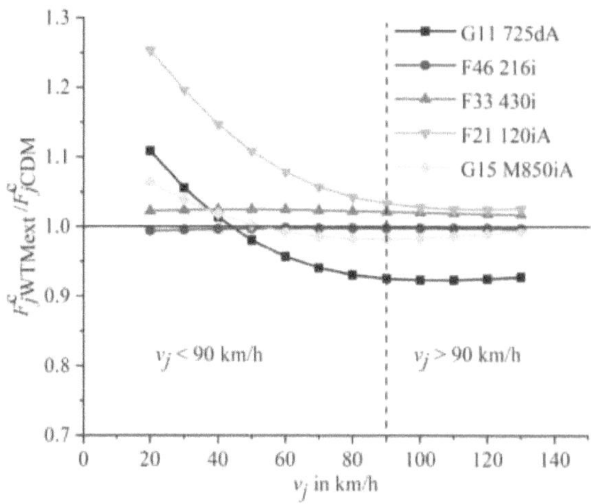

Figure 4.22: Ratios of the road load determined with the wind tunnel method extended $F_{j,\text{WTMext}}$ related to the coastdown method $F_{j,\text{CDM}}$ over the reference velocity points v_j for the five vehicles G11 725dA, F46 216i, F33 430i, F21 120iA and G15 M850iA.

total energy demand for the WLTC is transformed in about the last $252\,\text{s}$ ($\approx 5\,\text{min}$), which corresponds even to the last $14\,\%$ of the total cycle duration. Furthermore, these last time periods are part of the cycle sections with the highest average velocity: $96.9\,\text{km/h}$ for the section motorway 130 of the ARTEMIS European driving cycle (see Table 2.5) and $92\,\text{km/h}$ for the section extra-high of the WLTC (see Table 2.7). This clarifies the significant influence of the high velocity range on the absolute road load values.

Nevertheless, the difference in cycle energy demand for the F21 120iA is 0.8 percentage points higher for the WLTC than for the ARTEMIS European driving cycle (see Table 4.4), although the maximum deviation of the road load determined with the wind tunnel method extended, when compared to the coastdown method, is smaller than $3\,\%$ for the velocity range greater than $90\,\text{km/h}$. The reason can be found in the increase of the road load in the velocity range smaller than $90\,\text{km/h}$ for the wind tunnel method extended when compared to the coastdown method. The overprediction due to the wind tunnel method extended is about $25\,\%$ at the reference velocity point of $20\,\text{km/h}$ in this case. As can be seen in Table 2.8, the velocity share in the range $0\,\text{km/h} < v \leq 90\,\text{km/h}$ is about $71.8\,\%$ for the WLTC, whereby the part $\leq 50\,\text{km/h}$ accounts for $45.7\,\%$. In contrast, for the ARTEMIS European driving cycle, the part with velocities $\leq 90\ \text{km/h}$ is only about $63.2\,\%$ with the share $0\,\text{km/h} < v \leq 50\,\text{km/h}$ of $37.8\,\%$. Furthermore, in Figure 4.24 the

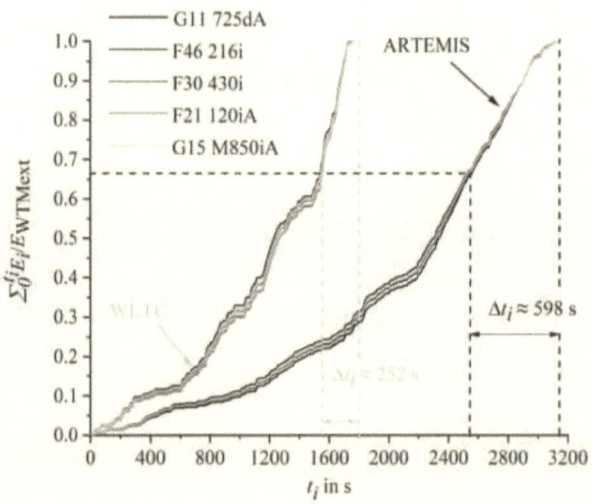

Figure 4.23: Energy demand for both cycles WLTC and ARTEMIS European driving cycle over time t_i related to the total energy demand of the cycles for the five vehicles G11 725dA, F46 216i, F33 430i, F21 120iA and G15 M850iA.

distributions of energy demand, which are transformed, depending on the three different velocity ranges 'low' ($0\,\mathrm{km/h} < v \leq 50\,\mathrm{km/h}$), 'medium' ($50\,\mathrm{km/h} < v \leq 90\,\mathrm{km/h}$) and 'high' ($v > 90\,\mathrm{km/h}$), are illustrated for the five vehicles and for both driving cycles. For the ARTEMIS European driving cycle it can be seen that approximately 70 % of the total energy demand is transformed in the high velocity range ($v > 90\,\mathrm{km/h}$). This is also the largest share for WLTC. However, it only accounts to about 40 %. Otherwise, the other two velocity ranges, 'low' and 'medium', have almost the same share of 30 % in the WLTC. In contrast, for the ARTEMIS European driving cycle, only 30 % of the total energy demand is transformed in the low and medium velocity range. This illustrates that the lower velocity sections in the WLTC have a higher influence on the calculated energy demand as compared to the ARTEMIS European driving cycle. This explains the higher value for the difference in cycle energy demand for the WLTC as compared to the ARTEMIS European driving cycle for the F21 120iA. The figures for the ratios of the road load determined with the wind tunnel method extended related to the coastdown method (see Figure C.1) and for the distributions of the energy demand transformed depending on the velocity ranges (see Figure C.2) for the eight vehicles used for the verification of the wind tunnel method extended are provided in the appendix .

Furthermore, for the vehicles with an automatic transmission, an increase of the curve progression with decreasing velocity is observed (see the test vehicles G11 725dA,

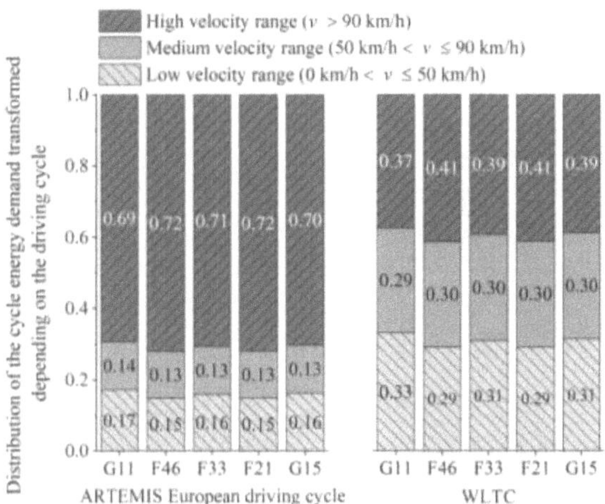

Figure 4.24: Distribution of the energy demand transformed depending on the velocity ranges 'low', 'medium' and 'high' for the five vehicles and for both driving cycles.

F21 120iA and G15 M850iA in Figure 4.22). This deviation between the wind tunnel method extended and the coastdown method is explained in the following. Using the coastdown method, the time corresponding to the reference velocity v_j is measured as the elapsed time t_{jAi}/t_{jBi} from the vehicle velocity $(v_j + 5\,^{\text{km}}/_{\text{h}})$ to $(v_j - 5\,^{\text{km}}/_{\text{h}})$. In addition to that, the coastdown runs are carried out in opposite directions (A and B) to eliminate influences of the wind and of the test track (see Figure 2.25). Afterwards, the coastdown times t_{jAi} in direction A and t_{jBi} in direction B are initially averaged separately (see Equations 2.43 and 2.44). The harmonic average Δt_j of these alternate coastdown time measurements Δt_{jA} in direction A and Δt_{jB} in direction B is calculated using Equation 2.42. Finally, the road load F_j (see Equation 2.36) is calculated using the coastdown time Δt_j, which is averaged in two steps. Additionally, it is assumed that the deceleration between the two velocity points $(v_j + 5^{\text{km}}/_{\text{h}})$ and $(v_j - 5^{\text{km}}/_{\text{h}})$ is constant (see Equation 2.37). In contrast, using the wind tunnel method extended, the road load part containing the drivetrain losses is directly measured in Newtons with the load cells of the flat belt dynamometer (see subsection 2.3.2). To clarify the effect of these differences in the test procedures, the road load for each Coastdown Run (CR) is directly calculated with the measured coastdown times t_{jAi} in direction A and t_{jBi} in direction B. The resulting road load curves for the single coastdown runs $F_{j,\text{CR},A}$ in direction A (light grey dashed lines) and $F_{j,\text{CR},B}$ in direction B (light grey dotted lines) are plotted in Figure 4.25 in combination with the uncorrected road load $F_{j,\text{WTMext}}$ measured using the wind tunnel method extended (light grey line).

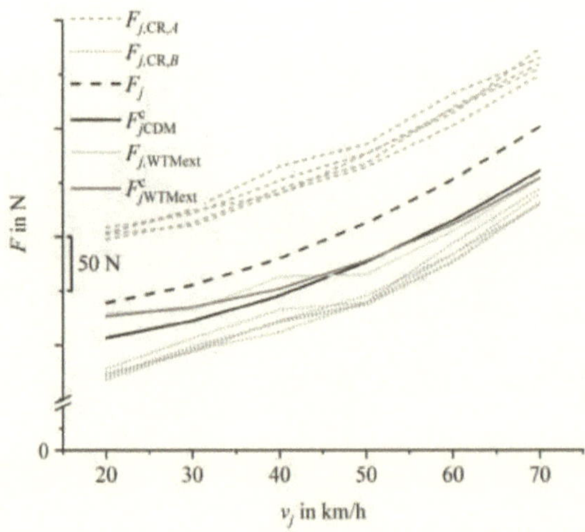

Figure 4.25: Influence of the gear shifting of the 8HP automatic transmission on the road load determined using the coastdown method and the wind tunnel method extended in the velocity range from $20\,\mathrm{km/h}$ to $70\,\mathrm{km/h}$. Illustrated are the single coastdown runs in direction A $F_{j,\mathrm{CR},A}$ (light grey dashed lines), the single coastdown runs in direction B $F_{j,\mathrm{CR},B}$ (light grey dotted lines), the uncorrected road load measured using the coastdown method F_j (black dashed line), the corrected road load corresponding to reference conditions $F^{\mathrm{c}}_{j,\mathrm{CDM}}$ (black line), the uncorrected road load measured using the wind tunnel method extended $F_{j,\mathrm{WTMext}}$ (light grey line) and the corrected road load measured using the wind tunnel method extended $F^{\mathrm{c}}_{j,\mathrm{WTMext}}$ (dark grey line).

In all three cases, the influence of the automatic transmission on the road load can be observed. In the next step, the road load F_j (black dashed line) calculated according to Equation 2.36, which uses the two times averaged coastdown time Δt_j, and the corrected road load corresponding to reference conditions $F^{\mathrm{c}}_{j,\mathrm{CDM}}$ (black line) calculated using the road load coefficients (see Equation 2.50) are plotted. It is obvious that due to the averaging steps, the influence of the gear shifting of the automatic transmission in the velocity range from $20\,\mathrm{km/h}$ to $50\,\mathrm{km/h}$ is not visible anymore in the road loads determined using the coastdown method. In contrast, using the road load determination according to the wind tunnel method extended, the influence of the gear shifting is only averaged out in the last step, when the road load is calculated using the road load coefficients ($F^{\mathrm{c}}_{j,\mathrm{WTMext}}$) (dark grey line). This can result in a higher road load in the lower velocity range for the wind tunnel method extended as compared to the coastdown method, as illustrated for the

G15 M850iA in Figure 4.25. In this figure, only the velocity range from 20 $^{km}/_h$ to 70 $^{km}/_h$ is illustrated. The velocity range from 80 $^{km}/_h$ to 130 $^{km}/_h$ is shown in Figure C.3 in the appendix.

4.2.4 Summary

- An optimum correction factor K_{ext} for the wind tunnel method extended is determined on the basis of five different vehicles. It has a value of 1.34.

- The wind tunnel method extended with $K_{ext} = 1.34$ is verified with further eight vehicles.

- For all vehicles, the difference in cycle energy demand ϵ between the coastdown method and the wind tunnel method extended lie in the range $\pm 5\,\%$, which is also an acceptance criterion for the wind tunnel method according to GTR No. 15 [10].

- One third of the total energy demand is transformed in about the last 19 % of the total cycle duration of the ARTEMIS European driving cycle and about the last 14 % of the WLTC. Additionally, these time periods are part of the cycle sections with the highest average velocity (96.9 $^{km}/_h$ for the ARTEMIS European driving cycle and 92 $^{km}/_h$ for WLTC).

- The difference in cycle energy demand depends on the driving cycle which is chosen for the calculation. Although it is shown that more than one third of the total energy demand is transformed in the high velocity range of WLTC, also the low and medium velocity ranges have a significant influence on the result. In both velocity ranges, one third of the total energy demand is transformed, respectively. In contrast, for the ARTEMIS European driving cycle, the same absolute differences in road load in the low and medium velocity ranges do not have such a high impact on the result of the calculated difference in cycle energy ϵ, as only 30 % of the total energy demand in transformed in these velocity ranges in total. The remaining part of approximately 70 % is converted in the high velocity range.

Note: The wind tunnel method extended differs from the wind tunnel method described in GTR No. 15. The wind tunnel method extended is only used in the context of the present study.

4.3 AEROLAB method

In this section, the results for the road load determination using the AEROLAB Method (AM) are presented. The method is already introduced in subsection 3.1.3. First, the test procedure is discussed in detail and then the results of the AEROLAB method are compared to the road load determined by using the wind tunnel method extended and the coastdown method. For all results in this section, the test vehicle F46 216i is used.

The determination of the road load according to the AEROLAB method is divided into two parts. The road load of a vehicle including the aerodynamic drag $F_{j,\text{WT,woB}}$, but without the residual brake forces $F_{j,\text{Brake}}$ is determined in the AEROLAB wind tunnel of the BMW Group. As the AEROLAB wind tunnel is not able to do a driving simulation and has no exhaust extraction system, the vehicle cannot accelerate by itself and decelerate afterwards. Therefore, it is ensured that the results of the AEROLAB wind tunnel measurements do not comprise any residual brake forces. The missing residual brake forces $F_{j,\text{Brake}}$ are measured separately at the flat belt dynamometer. A schematic representation of the AEROLAB method is given in Figure 3.10.

In Figure 4.26 the measurement quality of the AEROLAB method is shown. The average of AEROLAB wind tunnel measurements $F_{j,\text{WT,woB}}$ (black line with squares) conducted over six months is shown. The corresponding standard deviation (black open squares) is plotted over the reference velocity points v_j using the right axis and is about 5 N on average. Additionally, the average of six residual brake force measurements $F_{j,\text{Brake}}$ (grey line with dots) and its standard deviations (grey open circles) are also plotted in Figure 4.26. As the standard deviation of the residual brake force measurements is exhibiting 5 N over the complete velocity range, the calculated residual brake forces are verified in the following.

For the verification of the determined residual brake forces $F_{j,\text{Brake}}$, illustrated in Figure 4.26, the road load with residual brake forces $F_{j,\text{Dyno}}$ are measured at the flat belt dynamometer using the test procedure shown in Figure 2.28. Additionally, the road load without residual brake forces $F_{j,\text{Dyno,woB}}$ are also measured at the flat belt dynamometer, using the test procedure SV ALT woB (see in Figure A.1 with the stabilization time of 4 seconds). In the next step, the sum (grey line with triangles) of the road loads without residual brake forces $F_{j,\text{Dyno,woB}}$ and the residual brake forces $F_{j,\text{Brake}}$ are compared in Figure 4.27 with the road loads $F_{j,\text{Dyno}}$ (black line with squares). The difference ΔF_j (black open circles) between these two road load curves is plotted in Figure 4.27 using the right axis. The dashed lines mark the total accuracy of the four load cells of the flat belt dynamometer ($\pm 1.2\,\text{N}^{12}$).

It is becoming clear that the sum of the road load without residual brake forces $F_{j,\text{Dyno,woB}}$ and the residual brake forces $F_{j,\text{Brake}}$ determined separately (grey line with triangles) does not differ by more than about 3.5 N from the road load measurements with included residual brake forces $F_{j,\text{Dyno}}$ (black line with squares). In the velocity range from $80\,^{\text{km}}/_{\text{h}}$ to $130\,^{\text{km}}/_{\text{h}}$,

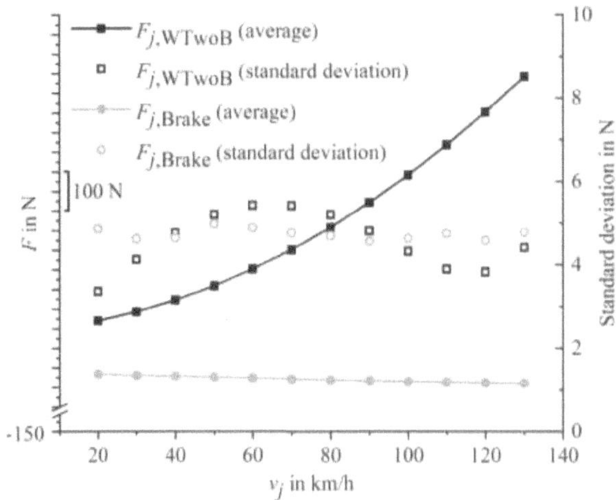

Figure 4.26: Average of AEROLAB wind tunnel measurements $F_{j,\mathrm{WTwoB}}$ (black line with squares) with standard deviation (black open squares) and the average of residual brake force measurements $F_{j,\mathrm{Brake}}$ at the flat belt dynamometer (grey line with dots) with the standard deviation (grey open cirlces) plotted over the reference velocity points v_j.

the maximum absolute deviation is about 2 N. Considering that the accuracy of the load cells is in total $\pm 1.2\,\mathrm{N}^{12}$, these differences are comparatively small. As a consequence, the residual brake forces $F_{j,\mathrm{Brake}}$ determined separately are verified for this vehicle and can be used for the determination of the road load according to the AEROLAB method.

Furthermore, the standard deviations for the flat belt dynamometer measurements with residual brake forces $F_{j,\mathrm{Dyno}}$ (black squares) and without residual brake forces $F_{j,\mathrm{Dyno,woB}}$ (grey triangles) are plotted over the reference velocity points v_j in Figure 4.28. It shows that the measurements for the road load without residual brake forces $F_{j,\mathrm{Dyno,woB}}$ have only a standard deviation of about 1.4 N on average over the complete velocity range (grey open triangles). Therefore, they are close to the accuracy limit of the flat belt dynamometer of $1.2\,\mathrm{N}^{12}$. In comparison, the standard deviation for the measurements with braking phase $F_{j,\mathrm{Dyno}}$ rises with increasing velocity from 2.4 N up to about 5.2 N, which clearly demonstrates that in this case the residual brake forces are mainly responsible for uncertainties due to the road load determination according to both methods, the AEROLAB method and the wind tunnel method extended.

Finally, the road load determined using the coastdown method (see subsection 3.1.1), the wind tunnel method extended (see subsection 3.1.2) and AEROLAB method (see subsection 3.1.3) are determined for the test vehicle F46 216i and then compared. In

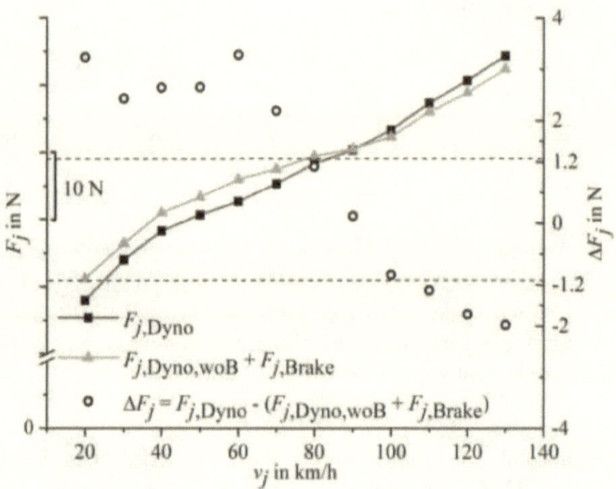

Figure 4.27: Road load with residual brake forces $F_{j,\text{Dyno}}$ (blake line with squares) as well as the sum (grey line with triangles) of the road load without residual brake forces $F_{j,\text{Dyno,woB}}$ and the separately determined residual brake forces $F_{j,\text{Brake}}$ and the difference ΔF_j (black open circles) between these two road load curves over the reference velocity points v_j.

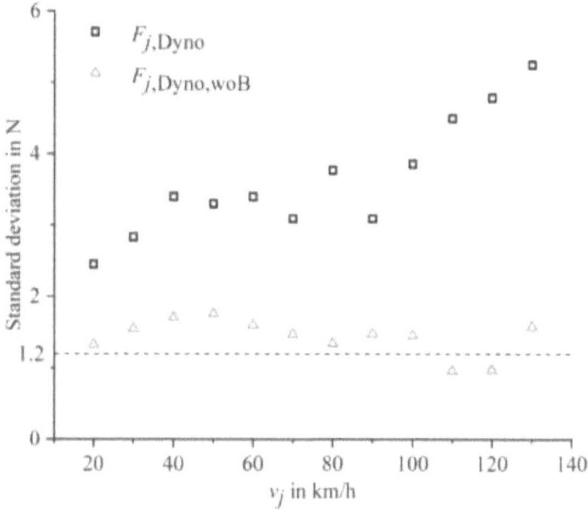

Figure 4.28: Standard deviation of the road load with residual brake forces $F_{j,\mathrm{Dyno}}$ (black squares) and without residual brake forces $F_{j,\mathrm{Dyno,woB}}$ (grey triangles) measured using the flat belt dynamometer plotted over the reference velocity points v_j.

Figure 4.29, the resulting road load for the AEROLAB method (grey line with dots), for the wind tunnel method extended (dark grey line with triangles) and for the coastdown method (black line with squares), normalized to the coastdown method, are illustrated over the reference velocity points v_j. The results for the wind tunnel method extended are identical to the results already presented in Figure 4.22.

It becomes clear that the road load determined using the wind tunnel method extended differs by a maximum of about 0.6 % at the reference velocity point of $20\,\mathrm{km/h}$. In comparison, the road load determined with the AEROLAB method is also underestimated over the complete velocity range. However, the determined road load differs by about a maximum of 3 % at $30\,\mathrm{km/h}$. The minimum difference of about 1 % occurs at $130\,\mathrm{km/h}$ for both methods. Finally, the differences in cycle energy demand between the coastdown method and the wind tunnel method extended $\epsilon_{\mathrm{WTMext}}$ and between the coastdown method and the AEROLAB method ϵ_{AM} are given in Table 4.6 for both driving cycles (WLTC and ARTEMIS European driving cycle).

It can be seen that for both the wind tunnel method extended and the AEROLAB method, the differences in cycle energy demand are smaller than $\pm 5\,\%$, which is also an acceptance criterion according to GTR No. 15 [10]. The maximum difference of -0.9 % can be found for the AEROLAB method using the WLTC.

However, it has to be considered that for the road load determined with the wind tunnel

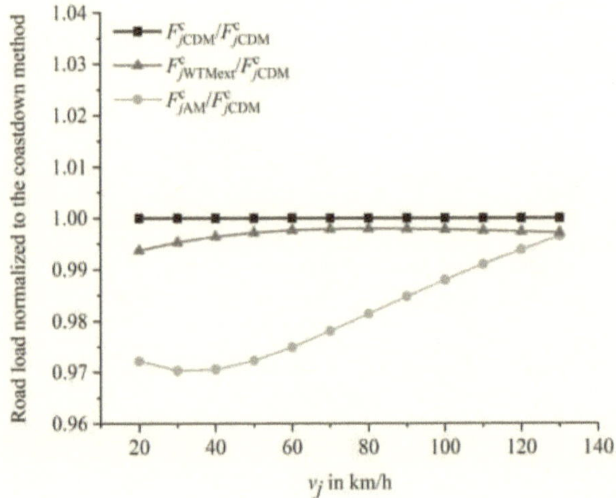

Figure 4.29: Road load determined using the AEROLAB method (light grey line with dots), the wind tunnel method extended (dark grey line with triangles) and the coastdown method (black line with squares) normalized to the coastdown method plotted over the reference velocity points v_j.

Table 4.6: Differences in energy demand between the wind tunnel method extended with $K_{\text{ext}} = 1.34$ and the coastdown method ϵ_{WTMext} and between the AEROLAB method and the coastdown method ϵ_{AM} for the test vehicle F46 216i

Difference in cycle energy demand in %	WLTC	ARTEMIS
ϵ_{WTMext}	-0.2	-0.2
ϵ_{AM}	-0.9	-0.7

method extended, the rolling resistance is corrected by 34 % to reach the road load level comparable to the coastdown method. In contrast, besides the correction to reference conditions, there is no further correction term which is applied on the road load determined using the AEROLAB method (compare subsection 3.1.3). Therefore, in the following subsections, different factors are discussed which influence the road load measured in the AEROLAB wind tunnel.

4.3.1 Influencing factors on the road load determined using the AEROLAB method

The investigated influencing factors are sorted according to the following categories:

- Differences in the aerodynamic drag

- Differences in the test procedures

- Differences in the rolling resistance and the drivetrain losses

Differences in the aerodynamic drag

In the following subsection, several influencing factors referenced to the aerodynamic drag are discussed:

- Ride height change and wind velocity

- Additional aerodynamic drag due to the vehicle fixation system used in the AERO-LAB wind tunnel

- Additional wheel ventilation resistance

Afterwards, the impact of these influencing factors on the road load measured using the AEROLAB method is rated.

<u>Influence of ride height change and wind velocity on the aerodynamic drag</u>
In subsection 2.2.1, it is already mentioned that it is not possible to sum up individual resistances concerning the aerodynamic drag. Instead, the interaction between these has to be considered [16, 18]. In this subsection, the following influencing factors on the resulting aerodynamic drag are investigated:

- Ride height change

- Wind velocity

The aerodynamic drag coefficient for the wind tunnel method according to GTR No. 15 and also for the wind tunnel method extended is determined at a constant ride height and at a velocity of $140\,^{km}/_h$. The ride height is equal to the ride height of the standing vehicle, which is loaded with the vehicle test mass. In contrast, using the coastdown method and the AEROLAB method, the aerodynamic drag is determined depending on the wind velocity, whereby also the ride height changes. To evaluate this influence on the resulting road load, the ride height change compared to the ride height of the standing vehicle is measured during the coastdown method on a proving ground, during the AEROLAB method and also in the BMW Group wind tunnel. In Figure 4.30, the ride height changes for the wheels front left FL (a), front right FR (b), rear left RL (c) and rear right RR (d) are plotted over the reference velocity points v_j for the following methods:

- Averaged coastdown runs carried out in direction A on the test track (black lines)

- AEROLAB wind tunnel (black triangles)

- BMW Group wind tunnel, where, in this case, the ride height is not fixed but depends also on the wind velocity (grey dots)

Additionally, the ride height change for the BMW Group wind tunnel with fixed ride height (black squares) is plotted. This ride height change is 0 mm for all reference velocity points, since the vehicle is fixed by four rocker panel restraints at the constant ride height of the standing vehicle. The ride height changes for the coastdown runs carried out in direction B of the test track are not illustrated, since the road surface in this direction is not made of asphalt continuously, as for direction A. Instead, there are also concrete plates with gaps at the joints (see subsection 3.1.1). When the vehicle crosses these gaps, the measurement system for the ride height change (described in subsection 3.3.3 and illustrated in the Figures 3.11 and 3.12) is additionally jumping, which causes further measurement uncertainties.

In the illustrations for the front wheels in Figure 4.30 (a and b), it can be seen that during the coastdown runs from $140\,^{km}/_h$ to $20\,^{km}/_h$ the front of the vehicle is elevated by around 5 mm. In contrast, in the AEROLAB wind tunnel the vehicle is only elevated by around 1 mm. At the rear axle (c and d) a different behaviour is seen. Initially, the rear axle is elevated for all described test procedures. During the coastdown measurements the vehicle is then lowered by about 7.5 mm on the test track and by about 5 mm in the AEROLAB wind tunnel. Moreover, the ride height measurements in the BMW Group wind tunnel depending on the wind velocity have a similar curve characteristic as in the AEROLAB wind tunnel. The lower elevation at the rear axle in the BMW Group wind tunnel can be explained by the fixation system. In the BMW Group wind tunnel, the vehicle is fixed at the four rocker panels. In contrast, in the AEROLAB wind tunnel the vehicle is only fixed by the two front wheel hubs and thus the vehicle can more easily be elevated at the rear axle. In general, the results for the ride height changes are plausible, since down forces are measured at the front axle and lift is measured at the rear axle [60].

However, it can be seen that the ride height change for all four wheels would not go back to zero with a further decreasing velocity. The following three reasons are assumed for this:

- Tyre expanding with increasing velocity

- Increasing tyre diameter due to increasing tyre tread temperatures during the warm-up phase

- Hysteresis properties of the chassis

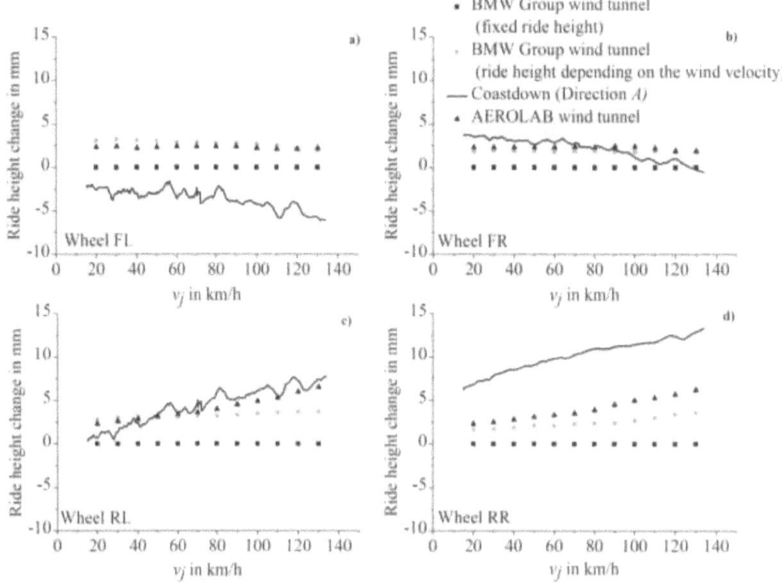

Figure 4.30: Ride height change Δc_w at the wheel front left FL (a), front right FR (b), rear left RL (c) and rear right RR (d) for the different methods (extended and adapted by permission from Isabell Vogeler: Different methods for road load determination in comparison: Wind tunnel, Wind tunnel method according to WLTP and Coastdown method [60], 2018).

In Figure 4.31, the distance between the rear infrared lasers (see Figures 3.11 and 3.12) and the ground of the test bench measured at the flat belt dynamometer during the test procedure SV ALT woB (see Figure A.1) is illustrated. The figure with the distance between the infrared lasers mounted at the vehicle front and the ground of the test bench is provided in the appendix (see Figure D.1). As the measurements are executed at the flat belt dynamometer, it can be excluded that the height change of this investigation is influenced by additional aerodynamic drag forces (compare the results given in subsection 4.1.1).

It can be seen that during the acceleration from $0\,\mathrm{km/h}$ to $144\,\mathrm{km/h}$ the distances between the lasers and the ground of the test bench increase. This is in accordance due to the phenomenon of tyre expanding with increasing velocity described in [77]. Furthermore, during the warm-up phase the tyre tread temperatures (see Figure D.2) and, therefore, also the tyre air temperatures increase during the warm-up phase[20]. Assuming the

---

[20]The correlation between the tyre tread temperature and the tyre air temperature in given in Figure 4.55 and Figure D.5.

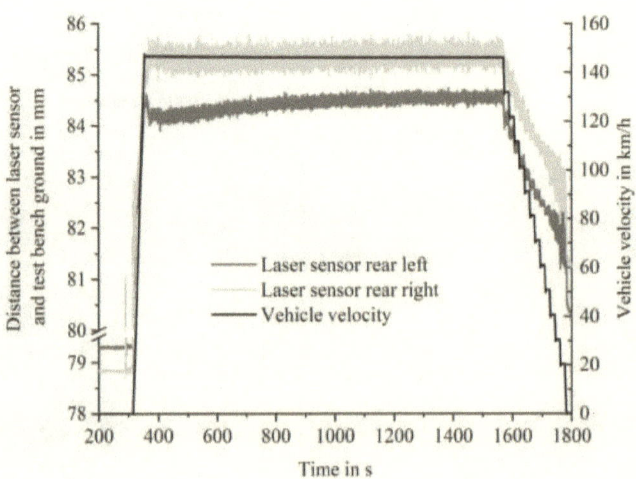

Figure 4.31: Distance between the laser rear left (dark grey line) and of the laser rear right (grey line) mounted at the vehicle and the ground of the test bench as well as the vehicle velocity (black line) during the test procedure SV ALT woB at the flat belt dynamometer.

Equation 2.12 for an ideal gas, this would result in a higher tyre pressure and, therefore, also in an increased tyre diameter. Furthermore, in Figure D.2 it can also be seen that the tyre tread temperatures decrease during the measurement. However, they are still higher than at the beginning of the test procedure. Considering this, it can be assumed that the tyre diameter is higher at the end than at the beginning of the test procedure. As third reason, it is assumed that there are hysteresis properties of the vehicle chassis, for example due to the spring-damper system (For more information about the spring-damper systems in vehicles see in [17].).

In the following, the influence of the ride height on the resulting aerodynamic drag coefficient is investigated. As already stated in subsection 2.2.1, a ride height change of about +5 to +7 mm for a BMW 3 Series sedan increases the aerodynamic drag coefficient by about 4 counts [24] and a ride height decrease of 30 mm for a AUDI Q7 decreases the coefficient by about 20 counts [20].

Therefore, the aerodynamic drag is measured again in the BMW Group wind tunnel, but now not with a fixed ride height. Instead, the ride heights depending on the wind velocity are used, which are measured on the test track in direction *A* and in both wind tunnels (see Figure 4.30). These ride heights are adjusted in the BMW Group wind tunnel for each reference velocity point and, afterwards, the aerodynamic drag is measured. The differences between the aerodynamic drag $c_{w(v_j,\text{float})}$, depending on ride height and the wind velocity, and the aerodynamic drag $c_{w(140,\text{fixed})}$, measured at constant ride height and at a velocity of 140 km/h, are shown in Figure 4.32. The calculation of the differences is expressed in the following equation:

$$\Delta c_w = c_{w(v_j,\text{float})} - c_{w(140,\text{fixed})} \tag{4.5}$$

where:

Δc_w	is the difference of the aerodynamic drag coefficient between c_w measured at 140 km/h with fixed ride height and c_w measured with ride height depending on the wind velocity;
$c_{w(140,\text{fixed})}$	is the aerodynamic drag coefficient measured in the BMW Group wind tunnel at 140 km/h and fixed ride height;
$c_{w(v_j,\text{float})}$	is the aerodynamic drag coefficient measured in the BMW Group wind tunnel with ride height depending on the wind velocity.

The differences are determined using:

- Coastdown runs in direction A (black line with crosses)

- AEROLAB wind tunnel (black line with triangles)

- BMW Group wind tunnel (grey line with dots)

The determined aerodynamic drag coefficients at the reference velocity point of 140 km/h shown in Figure 4.32 are measured with the ride height of the standing vehicle at the beginning of the ride height measurements in each case. It shows that the aerodynamic drag deviates only by 1 count. This verifies that the initial conditions of the vehicle are nearly the same for all measurement methods.

In addition, it can be seen that all determined c_w values, depending on the wind velocity and on the ride height, are higher in the velocity range from 70 km/h to 130 km/h as compared to the $c_{w(140,\text{fixed})}$ value measured at a fixed ride height at a velocity of 140 km/h. The maximal deviation of 9 counts occurs at the velocity of 80 km/h for the ride height change measured in the AEROLAB wind tunnel. Furthermore, the aerodynamic drag is reduced with increasing velocity in the velocity range from 70 km/h to 140 km/h. However, in the lower velocity range from 20 km/h to 60 km/h, a scatter of the measurement results can be seen. But due to the quadratic dependence of the velocity, the differences in the measured c_w values only cause a maximum delta of 2.4 N for the calculated aerodynamic drag at 60 km/h.

Nevertheless, there is no correlation identifiable between the change in ride height and the aerodynamic drag coefficient. However, it can be seen that the characteristic of the ride height change in both wind tunnels is similar. In addition, the determined aerodynamic drag only differs in the higher velocity range by a maximum of 3 counts. On the other

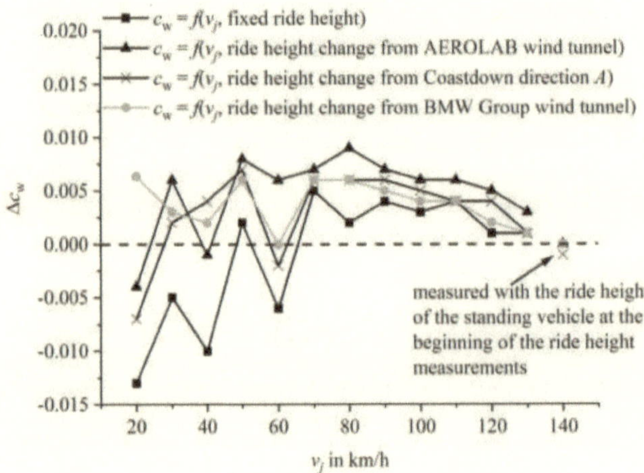

Figure 4.32: Differences between the aerodynamic drag measured with fixed ride height at the velocity of 140 km/h and the aerodynamic drag measured with ride height depending on the wind velocity of the coastdown runs in direction A (black line with crosses), in the AEROLAB wind tunnel (black line with triangles) and in the BMW Group wind tunnel both with fixed ride height (black line with squares) and with ride height depending on the applied ride height (grey line with dots) (extended and adapted by permission from Isabell Vogeler: Different methods for road load determination in comparison: Wind tunnel, Wind tunnel method according to WLTP and Coastdown method [60], 2018).

hand, the ride height changes during the coastdown runs show the same characteristics, but they have different absolute levels (see Figure 4.32). At the same time, the change in aerodynamic drag is similar to the measurements with the ride height changes from both wind tunnels. It can be concluded that, besides the ride height change, there must be another important effect, which has more influence on the aerodynamic drag.

As already stated in subsection 2.2.1 and illustrated in Figure 2.4, the effect of an aerodynamic drag reduction with increasing wind velocity is known. For the investigated vehicles (BMW Series 3 sedan, BMW Series 7 sedan and Rolls Royce Ghost) the aerodynamic drag at $80 \,\mathrm{km/h}$ is 5 counts higher as compared to the value at $140 \,\mathrm{km/h}$ (see Figure 2.4) [25].

On this account, the change in aerodynamic drag for the velocity range of $20 \,\mathrm{km/h}$ to $130 \,\mathrm{km/h}$ is measured in the BMW Group wind tunnel, but now with the fixed ride height of the standing vehicle. The difference as compared to the aerodynamic drag, which is determined with the same ride height at $140 \,\mathrm{km/h}$, is also illustrated in Figure 4.32 (black line and squares). It can be seen that the aerodynamic drag generally decreases with increasing wind velocities in the range from $70 \,\mathrm{km/h}$ to $130 \,\mathrm{km/h}$ [60], as is the case for the other measurement with changing ride height. This corresponds to the effect, which is illustrated in Figure 2.4 and is described in the Weber's work [25]. Furthermore, it shows that the absolute level of the coefficients are lower as compared to the measurements with additional ride height changes depending on the wind velocity.

With these results it is pointed out that the aerodynamic drag measured with the AEROLAB method has to be higher than when measured by the wind tunnel method according to GTR No. 15 and also by the wind tunnel method extended. The reason is the combination of measuring the aerodynamic drag depending on both the wind velocity and the ride height change for each reference velocity point. In contrast, the required aerodynamic drag coefficient for the wind tunnel method is only determined at a velocity of $140 \,\mathrm{km/h}$ and with a fixed ride height of the standing vehicle. This single coefficient is then used to determine the aerodynamic drag for all following reference velocity points from $130 \,\mathrm{km/h}$ to $20 \,\mathrm{km/h}$.

To evaluate the influence of the previously discussed aerodynamic effects on the road load determined using the AEROLAB method, the differences in the aerodynamic drag, illustrated in Figure 4.32, are separately subtracted from the AEROLAB method results. In Figure 4.33 only the curves for the AEROLAB method without the influence of the wind velocity but with fixed ride height (RH) $c_\mathrm{w} = f(v_j, \text{fixed RH})$ (light grey dashed line with triangle) and without the influence of the ride height change from the AEROLAB wind tunnel (WT) $c_\mathrm{w} = f(v_j, \text{RH of AEROLAB WT})$ (grey dashed line with dot) are shown, as these two influences exhibit the highest difference in the c_w values related to the c_w value, which is measured at fixed ride height at a velocity of $140 \,\mathrm{km/h}$ (see Figure 4.32). The other curves are provided in the appendix (see Figure D.3). The curves for the coastdown method (black line with squares), for the AEROLAB method (grey line with dots) and for the wind tunnel method extended (dark grey line with triangles) are the same as those shown in Figure 4.29.

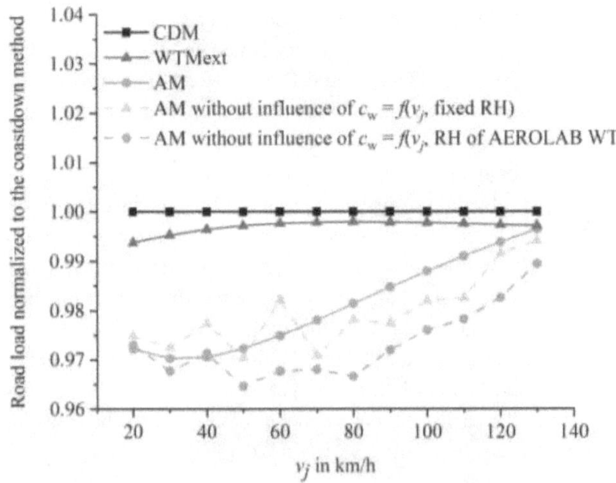

Figure 4.33: Difference in the determined road load of a vehicle between the wind tunnel method extended (dark grey line with triangles), the AEROLAB method with (grey line with dots) and without the influence of the aerodynamic drag depending on the wind velocity but fixed ride height (light grey dashed line with triangles) and without the influence of the aerodynamic drag depending on the wind velocity and the ride height change from the AEROLAB wind tunnel (grey dashed line with dots) normalized to the coastdown method (black line with squares).

It can be concluded that both the aerodynamic drag depending only on the wind velocity $c_w = f(v_j,$ fixed RH) and the aerodynamic drag depending on both the wind velocity and the ride height change in the AEROLAB wind tunnel $c_w = f(v_j,$ RH of AEROLAB WT) have a significant influence on the determined road load, especially in the higher velocity range. The maximum deviation between the AEROLAB method (grey line with dot) and the AEROLAB method without the influence of $c_w = f(v_j,$ RH of AEROLAB WT) (grey dashed line with dot) is 1.5 percentage points at the velocity point of 80 km/h. The curve for the AEROLAB method without the influence of $c_w = f(v_j,$ fixed RH) (light grey dashed line with triangles) lies between the other two curves in the velocity range from 70 km/h to 130 km/h as already expected from the results presented in Figure 4.32. Furthermore, it can be seen that the differences in the c_w values in the low velocity range from $20\,\mathrm{km/h}$ to $60\,\mathrm{km/h}$ (see Figure 4.32) do not have a significant influence on the road load (see Figure 4.33). The reason for this is that the resulting aerodynamic drag calculated from the c_w values is relatively low in this velocity range, due to its quadratic dependence on the velocity.

Influence of the vehicle fixation system on the aerodynamic drag
The aerodynamic drag of the vehicle fixation system components marked with b in Figure 3.7 and in Figure 3.8 are also measured with the integrated load cells (marked with c). Consequently, this influence is estimated with the aid of a numerical simulation, conducted by the Aerodynamics Department of the BMW Group. They state that the vehicle fixation system has in total an aerodynamic drag coefficient of 0.225 at a velocity of $140\,\mathrm{km/h}$[21], whereby it has to be considered that the frontal area of the vehicle fixation system is only about 2 % of the vehicle frontal area.

Influence of the wheel ventilation resistance
As already stated in [19] and [29], the wheel ventilation resistance F_Vent is an internal force in wind tunnels with a 5-belt system, such as the BMW Group wind tunnel. The aerodynamic drag coefficient for the wind tunnel method extended is measured in the BMW Group wind tunnel and, therefore, does not contain the wheel ventilation resistance component. Furthermore, in subsection 4.1.1 it is shown that there is no aerodynamic drag in the measurement result of the flat belt dynamometer. In contrast, the wheel ventilation resistance is included in the road load determined using the AEROLAB method. This leads to a higher road load as compared to the road load determined with the wind tunnel method stated in the GTR No. 15 or with the wind tunnel method extended.

In the following, the wheel ventilation resistance for the test vehicle F46 216i is estimated due to the results of Link [29]. As demonstrated in Figure 3.11, the rim design of the vehicle has ten spokes. Therefore, the wheel ventilation resistance coefficient c_Vent is estimated with a value of 0.0114 (compare to Figure 2.18).

At this point it has to be clarified that for the determination of the wheel ventilation resistance according to Link [29], the vehicle is fixed by the rocker panels in the BMW Group wind tunnel. In contrast, in the AEROLAB wind tunnel the vehicle is fixed at the front wheel hubs. Therefore, in reality there is a difference in the measured wheel ventilation due to these two different vehicle fixation systems.

Effect of the aerodynamic influences on the road load determined according to the AM
All influences, which are considered in the previous subsections, lead to a higher road load using the AEROLAB method as compared to a road load determined with the wind tunnel method or the wind tunnel method extended. The influences are separately subtracted from the road load of the AEROLAB method to see its effect on the road load. The results are shown in Figure 4.34:

---

[21] AERODYNAMICS DEPARTMENT OF THE BMW GROUP. Estimation of the influence of the vehicle fixation system of the AEROLAB wind tunnel on the aerodynamic drag: e-mail, 05.04.2019.

- Ride height change and wind velocity (grey dashed line with dots)

- Vehicle fixation system (light grey dashed line with open circles)

- Wheel ventilation resistance (grey dashed line with open circles)

The illustrated road loads for the coastdown method (black line with squares), for the AEROLAB method (grey line with dots) and for the wind tunnel method extended (dark grey line with triangles) are the same as those shown in Figure 4.29.

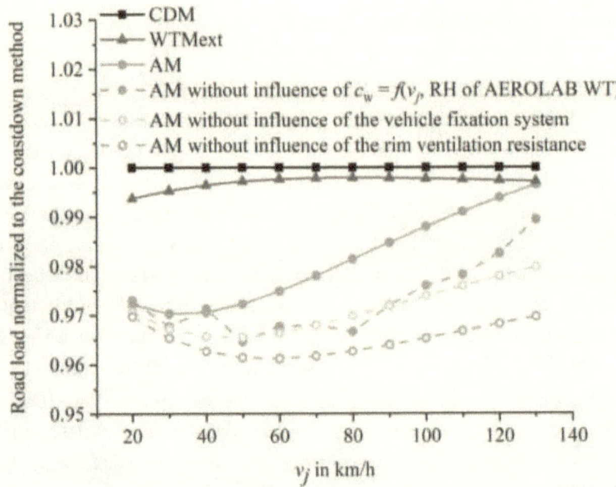

Figure 4.34: Differences in the determined road load of a vehicle between the coastdown method (black line with squares), the wind tunnel method extended (dark grey line with triangles), the AEROLAB method (grey line with dots) and the AEROLAB method without the differences in the aerodynamic drag (dashed lines) normalized to the coastdown method.

It can be seen that both the influence of the wind velocity and of the ride height change in the AEROLAB wind tunnel and the influence of the additional aerodynamic drag of the vehicle fixation system nearly have the same effect on the road load. The maximum deviation between these two curves is about 1 percentage point and occurs at the reference velocity point of 130 km/h. In this context, the wheel ventilation resistance has in this case the highest effect on the determined road load. If this influencing factor is subtracted, the road load normalized to the coastdown method is reduced by about 2.6 percentage points to a value of about 97 % at the reference velocity point of 130 km/h. In contrast, at a velocity of $20\,\text{km/h}$, the maximum difference between the AEROLAB method with and without the influence of the wheel ventilation resistance amounts to only about 0.3 percentage points. Additionally, it is pointed out that the effects on the road load for all three influencing factors increase with increasing velocity, due to the quadratic dependence of the aerodynamic drag.

Differences in the test procedures

In this subsection, the different influencing factors due to the test procedures used with the AEROLAB method and the wind tunnel method extended are discussed. As already mentioned, the standard test procedure according to the GTR No. 15 wind tunnel method had to be modified for the AEROLAB method. As there is no exhaust extraction system in the wind tunnel and the single-belt-rolling-road system in the wind tunnel is not able to realize a driving simulation, it is not allowed to measure with a running engine and the vehicle cannot accelerate and decelerate by itself in the wind tunnel. Therefore, the braking phase at the beginning is omitted. Furthermore, the stabilization time prior to each measurement point has to be increased from 4 seconds to 20 seconds for the reference velocity points ranging from $130\,\text{km/h}$ to $30\,\text{km/h}$, and to 30 seconds for the reference velocity point $20\,\text{km/h}$ (see Figure 3.9). This stabilization time is necessary, because the wind in the AEROLAB wind tunnel needs more time than the single-belt system to reach every reference velocity point. The effects of these influencing factors on the resulting road load determined with the AEROLAB method are discussed in detail in the following section.

<u>Influence of the external residual brake forces determination</u>

The omitted residual brake forces were separately determined using the flat belt dynamometer (see Figure 3.10). Afterwards, these brake forces were validated with further road load measurements at the flat belt dynamometer (see Figure 4.27). It can be seen that in the velocity range from $20\,\text{km/h}$ to $80\,\text{km/h}$, the separately determined residual brake forces lead to an overprediction of the road load using the AEROLAB method, whereby the maximum overprediction of about $3.3\,\text{N}$ occurs at $60\,\text{km/h}$. On the other hand, in the velocity range from $90\,\text{km/h}$ to $130\,\text{km/h}$, the separately determined residual brake forces lead to an underprediction of the road load, with a maximum value of about $-2\,\text{N}$ at $130\,\text{km/h}$.

<u>Influence of longer stabilization times during the measurement phase</u>

In this subsection, the influence of the stabilization time prior to each measurement point is investigated using the test procedures SV ALT woB as described in Figure 3.9 (increased stabilization time) and in Figure A.1 (standard stabilization time). In Figure 4.35, four different road load measurements conducted at the flat belt dynamometer are illustrated as a function of the reference velocity points v_j. The measurements with the standard stabilization time of $4\,\text{seconds}$ (black and grey lines with squares) are denoted 'Standard'. The measurements with increased stabilization times (black and grey lines with dots) are denoted 'Increased'. The average difference between these test procedures (black line with triangles) is also plotted using the right axis. It can be seen that for the most part the difference lies within the accuracy range of $\pm 1.2\,\text{N}$ of the load cells of the flat belt dynamometer[12] (dashed lines). The maximum difference, with a value of about $3\,\text{N}$, occurs at a velocity of $30\,\text{km/h}$. Therefore, it can be concluded that the increased stabilization times in the test procedure does not have a significant effect on the absolute road load determined with the AEROLAB method.

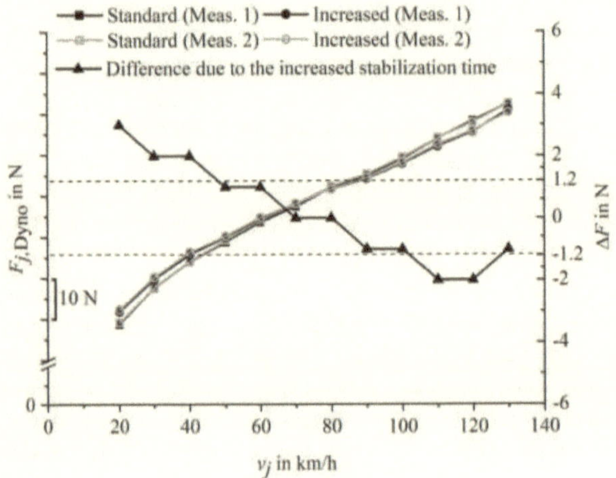

Figure 4.35: Difference between the road load determined using the test procedures SV ALT woB with a standard stabilization time of 4 seconds (black and grey lines with squares) and with an increased stabilization time up to 30 seconds (black and grey lines with dots) and the difference between these two test procedures (black line with triangles).

Nevertheless, the differences in the temperature profiles of the tyre treads and of the transmission oil are shown in Figure 4.36. The temperature profiles of the transmission oil (squares), the tyre tread front left FL (dots), FR front right (upward triangles), rear left RL (downward triangles) and rear right RR (diamonds) with the standard stabilization time of $4\,\text{s}$ (black lines) and the modified, increased time (grey lines) are illustrated over the reference velocity points v_j for two different measurements (filled and open symbols). It can be seen that the increased stabilization time has no effect on the transmission oil temperature. For both variations, the temperatures remain nearly constant over the total measurement phase. In contrast, the tyre tread temperatures fall more rapidly with decreasing velocity for the measurements with the modified stabilization time. However, this decrease, with about $3\,°\text{C}$, is not very significant. Therefore, no large effect on the resulting road load measured at the flat belt dynamometer is expected, which corresponds

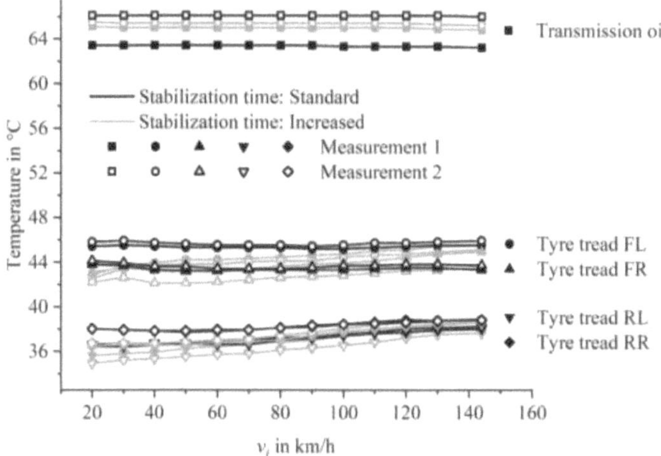

Figure 4.36: Influence of the stabilization time (Standard: black lines; Increased: grey lines) on the temperature profile of the transmission oil (squares), tyre tread FL (dots), tyre tread FR (upward triangles), tyre tread RL (downward triangles) and tyre tread RR (diamonds) over the reference velocity points v_j for two measurements (filled and open symbols) using the flat belt dynamometer, respectively.

to the results given in Figure 4.35.

With and without running engine

As there is no exhaust extraction system in the AEROLAB wind tunnel, it is not possible to determine the road load with a running engine. Thus, the difference in the road load measured with and without a running engine is evaluated in the following. As no difference in the rolling resistance is expected, only the difference in the drivetrain losses is investigated, determined using TOM (see subsection 3.3.4). Therefore, three measurements of the drivetrain losses $F_{j,\text{Drive}}$ with running engine (black line with squares) and without running engine (grey line with dots) are conducted with mounted TOM using the flat belt dynamometer. Moreover, the test procedure SV ALT woB with standard stabilization times described in Figure A.1 is used. The results with their corresponding standard deviation are illustrated in Figure 4.37. The difference ΔF (black line with triangles) between these two modifications is plotted using the right axis. The accuracy limits of $\pm 1.2\,\text{N}$ of the flat belt dynamometer load cells are again marked with dashed lines[12].

It can be seen that the drivetrain losses of a vehicle determined with the engine running is on average about $4.8\,\text{N}$ lower than the road load of a vehicle, which is measured without a

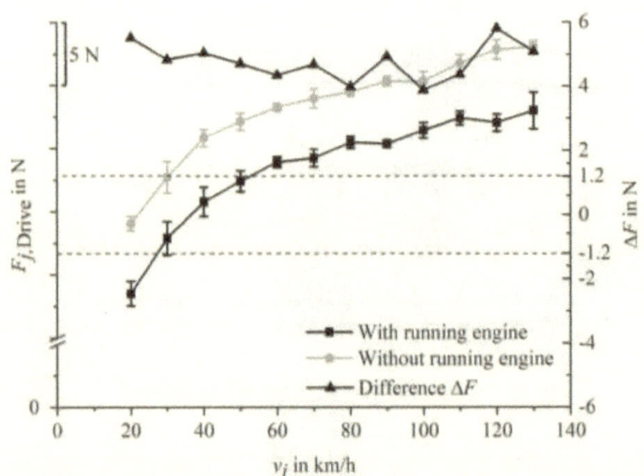

Figure 4.37: Drivetrain losses $F_{j,\mathrm{Drive}}$ measured using the flat belt dynamometer with mounted TOM and with running engine (black line and squares) and without a running engine (grey line with dots), with their corresponding standard deviations and the difference ΔF (black line with triangle), plotted over the reference velocity points v_j.

running engine. The maximum difference with a value of about 5.8 N occurs at a velocity of 20 km/h and the minimum difference of about 3.9 N at 100 km/h[22]. The reason for the difference in the road loads can be found in the temperature difference of the transmission oil, as shown in Figure 4.38. In this figure, the average transmission oil temperature of the measurements with a running engine (black line with squares) and without a running engine (grey line with dots), including their standard deviations, are plotted over the reference velocity points v_j. The temperature difference (black line with triangles) is shown using the right axis. The measurement point at 144 km/h corresponds to the temperature at the end of the warm-up phase.

Although the difference, with a maximum value of 3.4 °C at 20 km/h and a minimum value of about 2.5 °C at 144 km/h, is relatively low, the results correspond to the findings, reported in subsection 2.2.3, that the losses in the transmission gears decrease with increasing oil temperature. Furthermore, it can be seen that the temperature remains nearly constant over the total measurement phase, if the engine is running. If the engine is not running, the oil temperature generally decreases with decreasing velocity. However, it has to be pointed out that this decrease is only minor, approximately 1 °C.

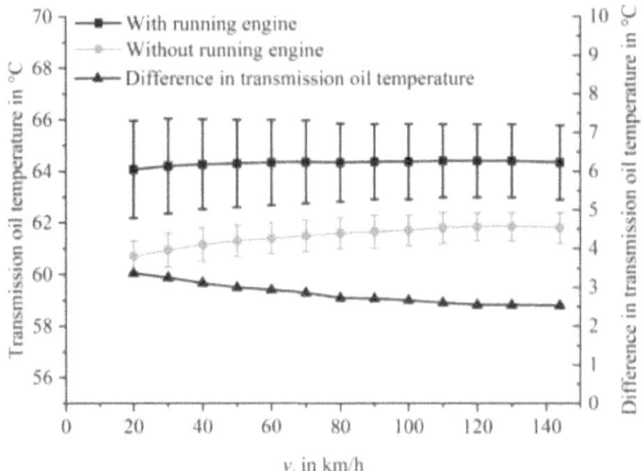

Figure 4.38: Transmission oil temperature during road load measurements using the flat belt dynamometer with a running engine (black line with squares) and without a running engine (grey line with dots), with their corresponding standard deviations and the difference between these two temperatures (black line with triangles), plotted over the velocity points.

Effect of the differences in the test procedures on the road load

The previously discussed influencing factors due to differences in the test procedures used for the AEROLAB method and the wind tunnel method extended are now evaluated according to their impact on the road load determined with the AEROLAB method. In Figure 4.39, the road load determined using the AEROLAB method without the following influencing factors is plotted over the reference velocity points v_j:

- Separately determination of the residual brake forces (grey dashed line with dots)

- Increased stabilization time in the AEROLAB wind tunnel (light grey dashed line with open circles)

- Lower transmission oil temperature due to a non-running engine (grey dashed line with open circles)

It can be seen that the difference due to the separate determination of the residual brake forces and due to the increased stabilization time have their main influence on the road load in the velocity range from $20\,\mathrm{km/h}$ to $60\,\mathrm{km/h}$. In contrast, if the difference in road load due to the lower transmission oil temperature is subtracted from the road load determined with the AEROLAB method, the road load is reduced by up to 2 percentage points.

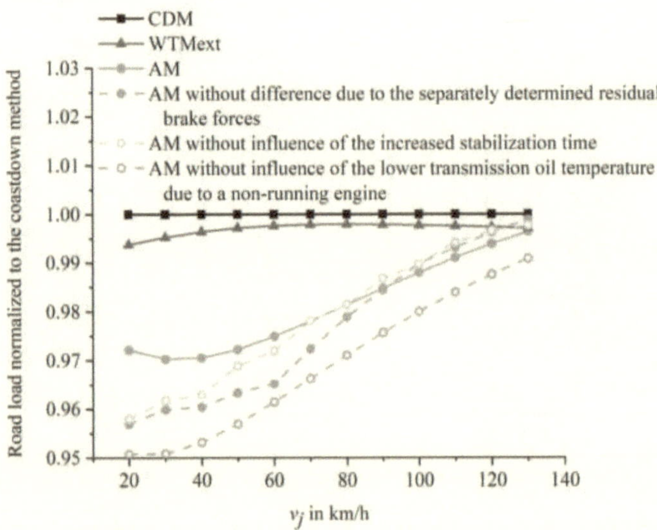

Figure 4.39: Differences in the determined road load of a vehicle between the coastdown method (black line with squares), the wind tunnel method extended (dark grey line with triangles), the AEROLAB method (grey line with dots) and the AEROLAB method without the differences in the test procedures (dashed lines), referenced to the coastdown method over the reference velocity points v_j.

Differences in rolling resistance and drivetrain losses

Not only differences in the aerodynamic drag determination and in the test procedures can be found but also constructive differences in the test benches, which have an influence on the measured rolling resistance and on the drivetrain losses. The following influencing factors are investigated:

- Vehicle fixation system

- Force measurement system

- Belt surface

- Transmission oil temperature and tyre tread temperatures

- Vehicle position

<u>Vehicle fixation system</u>

Using the flat belt dynamometer, the vehicle is usually fixed by the tow hooks at the front and at the rear of the vehicle, as can be seen in Figure 3.4 and in Figure 3.5. In contrast, in the AEROLAB wind tunnel, a vehicle fixation system is used, which is mounted at both front wheel hubs of the vehicle (see Figure 3.7). To investigate the differences in the determined road load due to these two vehicle fixation systems, the road load is measured at the flat belt dynamometer using the following two test setups:

- Standard fixation system of the flat belt dynamometer (tow hooks)

- Fixation system of the AEROLAB method (wheel hubs)

The test setup with the fixation system of the AEROLAB wind tunnel at the flat belt dynamometer is shown in Figure 4.40. In both cases, the force measurement is made with the four load cells of the flat belt dynamometer. The test procedure SV ALT woB (see Figure A.1) is used.

Figure 4.40: Test vehicle F46 216i fixed on the flat belt dynamometer with the vehicle fixation system from the AEROLAB wind tunnel, which uses both front wheel hubs.

In Figure 4.41, the averages of the road load measurements determined with the load cells of the flat belt dynamometer for the vehicle fixed by both front wheel hubs (black line with squares) and the vehicle fixed by its two tow hooks (grey line with dots) are illustrated over the reference velocity points v_j. The difference between these two curves (black line with triangles) is plotted using the right axis. Additionally, the accuracy limit of the load cells (black dashed lines) of $\pm 1.2\,\text{N}^{12}$ is shown.

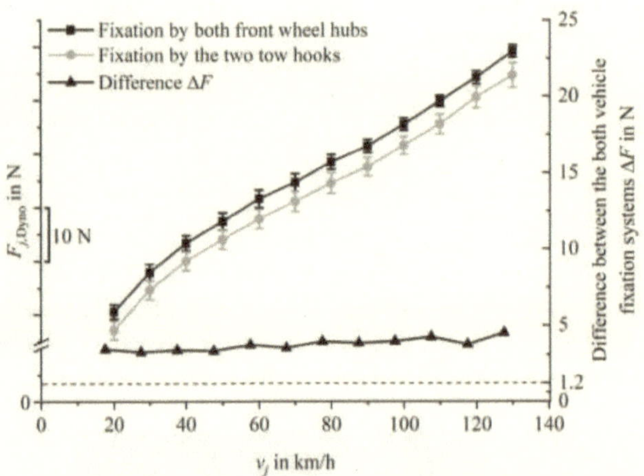

Figure 4.41: Road load measured with the load cells of the flat belt dynamometer for the vehicle fixed by both front wheel hubs (black line with squares) and for the vehicle fixed by its two tow hooks (grey line with dots).

The average difference in the road load between these two vehicle fixation systems is 3.7 N. The fixation system using the two front wheel hubs consists of further bearings (see Figure 3.7). It is assumed that the additional bearing losses are a reason for the higher road load measured using this vehicle fixation system.

Difference in the measured road load due to the different force measurement systems

However, not only the vehicle fixation system is different, but also the force measurement system. For the AEROLAB method two load cells are used, which are integrated into each side of the vehicle fixation system (see Figure 3.8). In contrast, with the flat belt dynamometer the load cells are located at the WDUs (compare Figure 3.4). Therefore, to compare these two force measurement systems, the test setup illustrated in Figure 4.40 is used. Moreover, the road load measurements are conducted using the test procedure SV ALT woB (see Figure A.1), whereby the road load is measured with both force measurement systems simultaneously. In Figure 4.42, the road load measured with the load cells integrated into the vehicle fixation system of the AEROLAB wind tunnel (black line with squares), the road load determined with the load cells of the flat belt dynamometer (grey line with dots) and the road load difference ΔF (black line with triangles) are illustrated over the reference velocity points v_j.

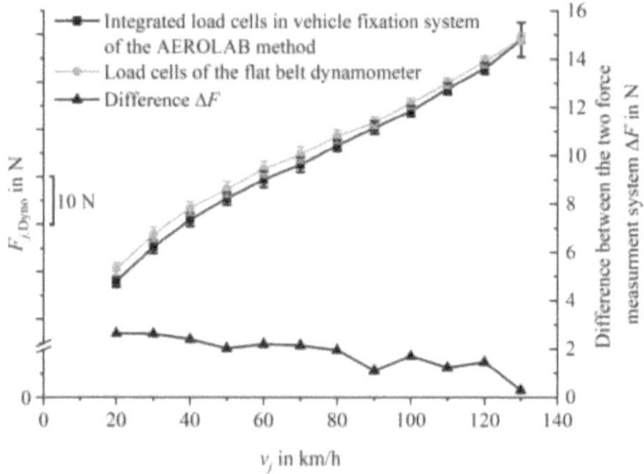

Figure 4.42: Road load measured with the load cells integrated in the vehicle fixation system (black line with squares), measured with the load cells of the flat belt dynamometer (grey line with dots) and the road load difference ΔF (black line with triangles) plotted over the reference velocity points v_j .

It can be seen that the maximum difference, with a value of about 2.6 N, occurs at a velocity of 20 km/h, whereby the difference decreases with increasing velocity up to a value of 0.3 N. Furthermore, it should be considered that the error bars for the measurements with both systems are overlapping and that each measurement system has a load cell accuracy of ±1.2 N in total (see subsections 3.1.2 and 3.1.3). Therefore, this difference is rated as a low influencing factor.

<u>Belt surface</u>

Another constructive difference between the flat belt dynamometer and the AEROLAB wind tunnel is the surface of the flat belts. At both test benches steel flat belts are used. However, at the flat belt dynamometer each flat belt is additionally coated with Safety Walk™. The difference in the resulting rolling resistance between a steel and a Safety Walk™ surface is not clearly defined. In Reimpell [39], it is stated that the rolling resistance on a steel surface is 5 % lower than on a Safety Walk™ surface. By contrast, in [40] a range between 2 % and 11 % is specified. However, in both cases the exact test procedures are not described. Considering this, the influence of the belt surface is also investigated at the tyre rear right (RR) of the test vehicle at the flat belt dynamometer. Due to the wide range of the impact provided in [39] and [40], the influence of the surface is investigated for three different wheel loads. For the measurements, the test procedure SV ALT woB (see Figure A.1) is used in combination with TOM (see subsection 3.3.4), to separate the flat belt dynamometer results into its two components, rolling resistance and drivetrain losses, using Equation 3.11. The effect of the increased wheel load on the rolling resistance is illustrated in Figure 4.43. In this figure, the rolling resistance at the wheel rear right with the standard wheel load defined by manufacturer specifications of about 3668.9 N (black line with squares), increased wheel load by about 26 % (grey line with dots) and by about 33 % (light grey line with triangles) referenced to the rolling resistance determined with standard wheel load is plotted.

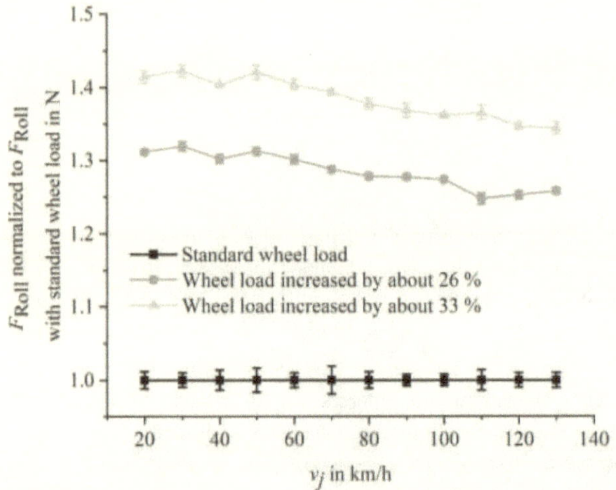

Figure 4.43: Rolling resistance F_{Roll} of the wheel rear right with standard wheel load (black line with squares), with an increased wheel load by about 26 % (grey line with dots) and by about 33 % (light grey line with triangles) referenced to the rolling resistance with the standard wheel load over the reference velocity point v_j.

It can be seen that there is a significant influence of the applied wheel load on the rolling resistance. With increasing wheel load, also the rolling resistance increases by about 26 % and 33 % on average.

Afterwards, the flat belt with the Safety Walk™ coating rear right (RR) is replaced by a steel flat belt (without coating), as shown in Figure 4.44. Furthermore, the rolling resistance is converted into the rolling resistance coefficient f_{r}, using Equation 2.15.

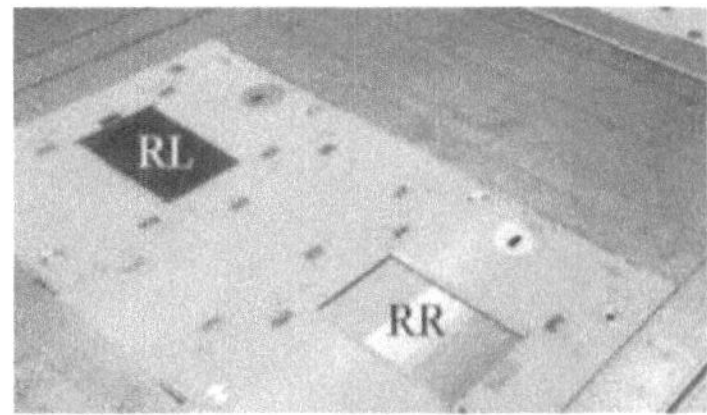

Figure 4.44: The flat belt dynamometer of the BMW Group, where the flat belt rear right (RR) with Safety Walk^{TM} coating is replaced by a steel flat belt without a coating.

In Figure 4.45, the ratios between the rolling resistance coefficient on the steel surface $f_{\text{r,Steel}}$ and on the Safety Walk^{TM} surface $f_{\text{r,Safety Walk}}$ is plotted over the reference velocity points v_j:

- Standard wheel load (black line with squares)

- Increased wheel load by about 26 % (dark grey line with dots)

- Increased wheel load by about 33 % (grey line with triangles)

The expected range given in [39] and [40] is illustrated as a hatched area in the background. However, there is neither a clear result for the ratios of the rolling resistance coefficient between the steel and the Safety Walk^{TM} surface nor a clear tendency due to the applied wheel load identifiable. It can be seen that the rolling resistance coefficient is reduced by about 2 % on the steel belt measured with the standard wheel load and with the wheel load, which is increased by about 33 %. In contrast, if the wheel load is only increased by about 26 %, the rolling resistance is reduced by about 5 %.

Taking into account that there is no clear result for the influence of the belt surface, a reduction of 3 % for the rolling resistance measured on a steel belt compared to the rolling resistance on a Safety Walk^{TM} surface is assumed in the following. This value is equal to the average of all three measurements illustrated in Figure 4.45 and lies in the range from 2 % to 11 % stated in [39] and [40].

In Figure 4.46 the temperature profiles of the tyre tread rear right for the standard wheel load (black line with squares), for the wheel load increased by about 26 % (grey line with dots) and for the wheel load increased by about 33 % (light grey line with triangles) on the Safety Walk^{TM} (filled symbols) and the steel surface (open symbols) are plotted over the reference velocity points v_j. In this case, the tyre tread temperature is measured with the infrared sensors of the flat belt dynamometer, which have an accuracy of $\pm 1\,^\circ\text{C}$[15].

It is pointed out that the tyre tread temperatures on the steel surface are always lower. This effect is also described in [19], where the influence of different tyre tread temperatures due to different surfaces on the measured rolling resistance is investigated. However, a more significant influence was observed. For example, it was shown that the tyre tread temperature of the same tyre with the same wheel load and inflation pressure is 65 °C on

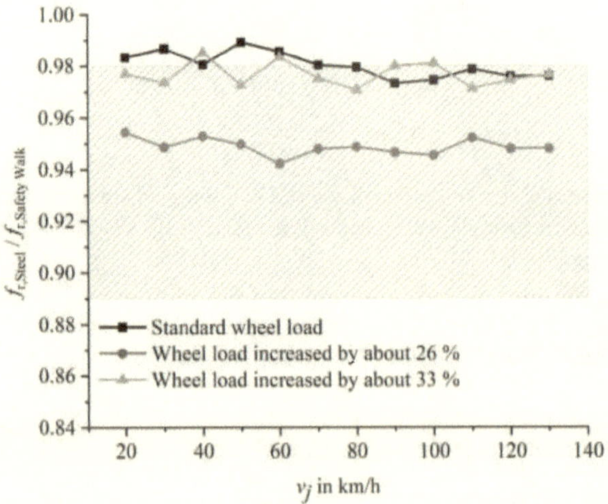

Figure 4.45: Ratio of the rolling resistance coefficients $f_{r,Steel}$ (steel surface) and $f_{r,Safety\ Walk}$ (Safety Walk™ surface) measured with a standard wheel load (black line with squares), with a wheel load increased by about 26% (dark grey line with dots) and increased by about 33% (grey line with triangles) as well as the expected range given in [39] and [40] illustrated as a hatched area.

an elastic band and 29 °C on a steel belt. In contrast, the maximum difference in this study is measured with standard wheel load and has a value of 1.5 °C. Therefore, no further influence is expected due to different tyre tread temperatures on the rolling resistance expected.

It can be concluded that the results found in this study correlate with the value range given in the literature [39] and [40] (see Figure 4.45). However, to build up a more reliable database, the influence of the belt surface on all four vehicles wheels and with more variation in the wheel load should be investigated in the future.

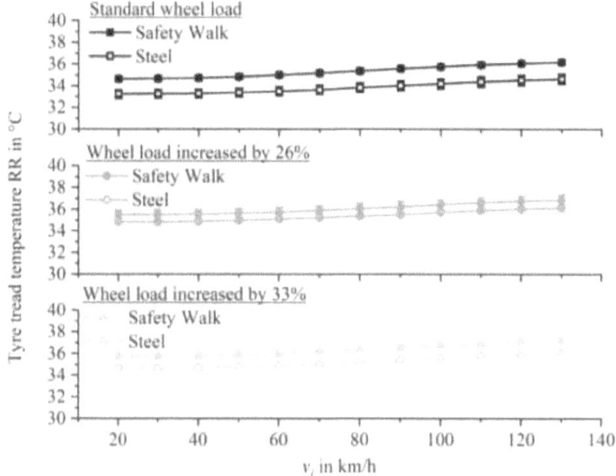

Figure 4.46: Temperature profiles of the tyre tread rear right RR with standard wheel load (black line with squares), with a wheel load increased by about 26 % (grey line with dots) and increased by about 33 % (light grey line with triangles) on a Safety Walk™ surface (filled symbols) and on a steel surface (open symbols) over the reference velocity points v_j.

Difference in tyre tread and transmission oil temperatures

Although the difference in the ambient temperature between the AEROLAB wind tunnel (black striped bars) and the flat belt dynamometer (black bars) is only about 3 °C, the differences in the tyre tread temperatures and the transmission oil temperature are even significantly lower, as can be seen in Figure 4.47.

It can be seen that the temperatures which are measured at the end of the warm-up phase in the AEROLAB wind tunnel are nearly at the same level as the temperatures measured at the flat belt dynamometer at an ambient temperature of 10 °C. In subsection 4.1.6, the influence of different ambient temperatures on the rolling resistance and on the drivetrain losses is already investigated. Therefore, it can be assumed that due to the lower vehicle temperatures in the AEROLAB wind tunnel, the proportions of the rolling resistance and the drivetrain losses determined using the AEROLAB method are higher as compared to the wind tunnel method extended [60]. At this point, it should be considered that in subsection 4.1.6 always the differences between the uncorrected road loads measured at the flat belt dynamometer for the different ambient temperatures are compared. However, the road load determined using the AEROLAB method is corrected

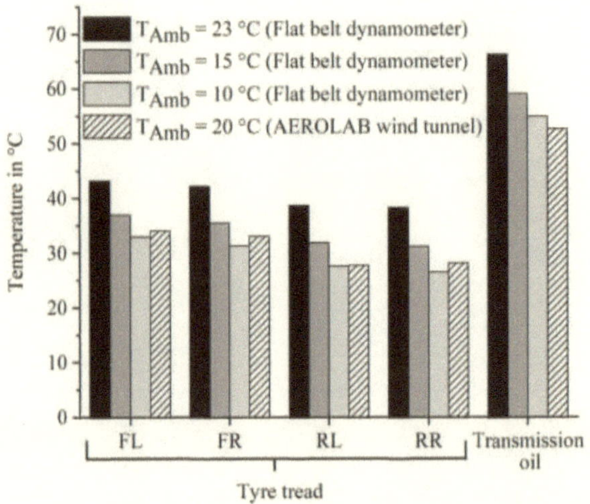

Figure 4.47: Tyre tread temperatures for all four tyres (FL = front left, FR = front right, RL = rear left and RR = rear right) and transmission oil temperatures for the different ambient temperatures 23 °C (black bars), 15 °C (grey bars), 10 °C (light grey bars) measured at the flat belt dynamometer and 20 °C (black striped bars) measured in the AEROLAB wind tunnel at the end of the warm-up phase (extended and adapted by permission from Isabell Vogeler: Different methods for road load determination in comparison: Wind tunnel, Wind tunnel method according to WLTP and Coastdown method [60], 2018).

to the reference temperature of 20 °C. Therefore, to evaluate the influence of the differences in tyre tread temperatures between the flat belt dynamometer and the AEROLAB wind tunnel, the resulting difference in rolling resistance ΔF_{Roll} is calculated as follows:

$$\Delta F_{\text{Roll}} = F_{\text{Roll,10,uncorr}} - F_{\text{Roll,23,corr}} \tag{4.6}$$

where:

ΔF_{Roll}	is the difference in rolling resistance measured at the flat belt dynamometer and measured in the AEROLAB wind tunnel in N;
$F_{\text{Roll,10,uncorr}}$	is the rolling resistance measured at the flat belt dynamometer at an ambient temperature of 10 °C in N;
$F_{\text{Roll,23,corr}}$	is the rolling resistance measured at the flat belt dynamometer at an ambient temperature of 23 °C, but corrected to the reference temperature of 20 °C using the Equations 2.27 and 2.46, in N.

Therefore, it can be assumed that the lower tyre tread temperatures and transmission oil temperature result in an increase of the rolling resistance of about 10 % and in an increase of the drivetrain losses of about 11 % on average.

So far, only the temperatures at the end of the warm-up phase are compared. In the following, the temperature profiles during the measurement phase in the AEROLAB wind tunnel and at the flat belt dynamometer are analyzed. In Figure 4.48, the tyre tread temperature profile for the tyres at the front axle, in Figure 4.49 for the tyres at the rear axle and in Figure 4.50 for the transmission oil are plotted over the reference velocity points v_j. In each figure the profiles at the flat belt dynamometer (filled symbols) with an ambient temperature of 23 °C for the left tyre (black line with squares) and the right tyre (grey line with dots) are compared with the profiles in the AEROLAB wind tunnel (open symbols) at an ambient temperature of 20 °C.

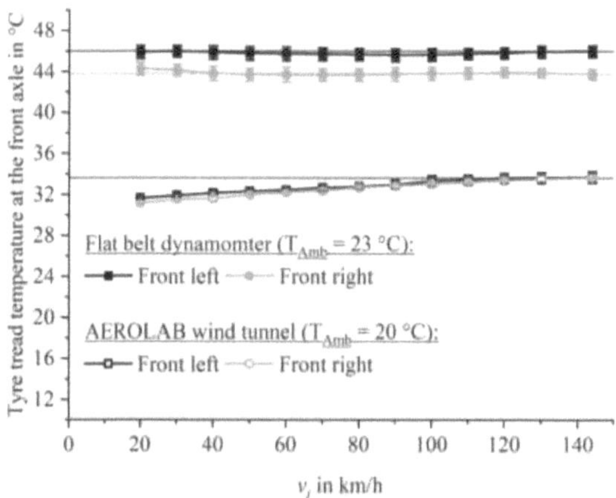

Figure 4.48: Tyre tread temperature profiles at the flat belt dynamometer (filled symbols) with an ambient temperature of 23 °C for the tyre front left (black line with squares) and the tyre front right (grey line with dots) are compared with the profiles in the AEROLAB wind tunnel (open symbols) at an ambient temperature of 20 °C at the end of the warm-up phase ($v_j = 144\,^{\text{km}}/_{\text{h}}$) and at the reference velocity points v_j.

It can be seen that the tyre tread temperature of the front wheels nearly remain constant over the measurement phase at the flat belt dynamometer (see Figure 4.48). In contrast, the front tyres in the AEROLAB wind tunnel cool down by about maximum 2.5 °C. At the rear axle (see Figure 4.49), a slight decrease can be discerned for both tyres in both test benches. However, this difference has only a maximum value of about 1.6 °C, taking into account that the maximum standard deviation is only about 0.6 °C.

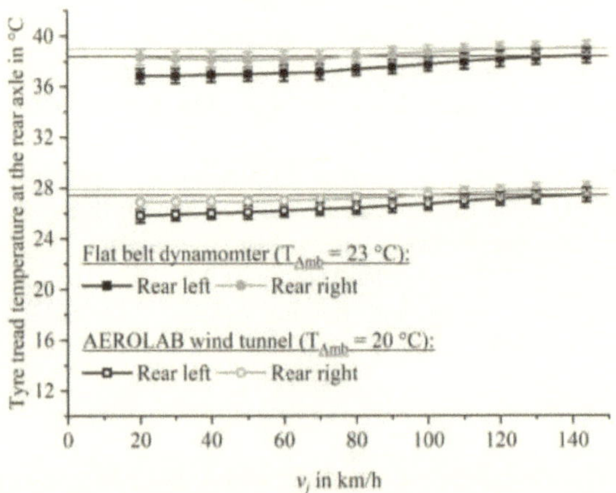

Figure 4.49: Tyre tread temperature profiles at the flat belt dynamometer (filled symbols) with an ambient temperature of 23 °C for the tyre rear left (black line with squares) and the rear right (grey line with dots) are compared with the profiles in the AEROLAB wind tunnel (open symbols) at an ambient temperature of 20 °C at the end of the warm-up phase ($v_j = 144\,\mathrm{km/h}$) and at the reference velocity points v_j.

For the transmission oil temperature a clear difference between the test benches is visible, which can be seen in Figure 4.50. The transmission oil temperature nearly remains constant during the measurement phase at the flat belt dynamometer (black line with squares) at an ambient temperature of 23 °C and also for the reduced ambient temperatures 15 °C and 10 °C. By contrast, the transmission oil temperature decreases by about 3.6 °C in the AEROLAB wind tunnel at an ambient temperature of 20 °C (black line with open squares). Although the temperature differences for the tyres and for the transmission oil are relatively low, a different cooling behaviour is visible at the flat belt dynamometer and in the AEROLAB wind tunnel. Since the transmission oil temperatures are nearly constant for all three ambient temperatures using the flat belt dynamometer, it is assumed that the vehicle temperatures in a test bench are not only a function of the ambient temperature. In the previous subsection, it has already been shown that both the increased stabilization time (see Figure 4.36) and the belt surface (see Figure 4.46) have an influence on the tyre tread temperature, whereby the maximum difference is about 1.6 °C. Moreover, there is a temperature difference between the tyre temperatures in the AEROLAB wind tunnel and at the flat belt dynamometer of about 10 °C, although the ambient temperature difference is only about 3 °C. Additionally, the transmission oil temperature differs also by about 13 °C. In a previous subsection, it was also shown that there is a difference in the transmission oil temperature, dependent on whether the engine is running or not running.

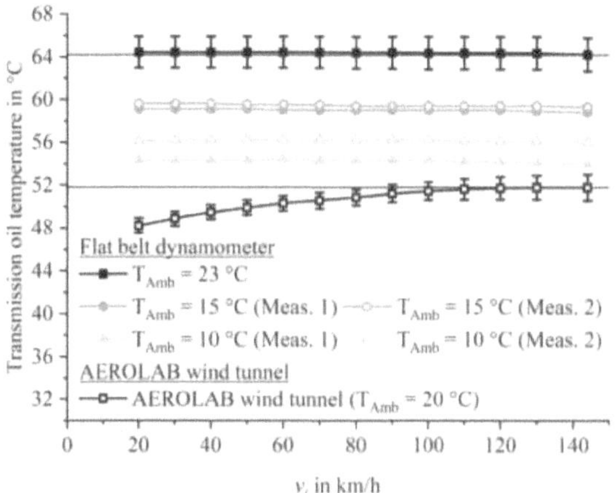

Figure 4.50: Temperature profile for the transmission oil at the flat belt dynamometer for the different ambient temperature of 23 °C (black line with square), 15 °C (grey line with dots) and 10 °C (light grey line with triangles) compared with the profile in the AEROLAB wind tunnel at an ambient temperature of 20 °C (black line with open squares) at the end of the warm-up phase (v_j = 144 $^{km}/_h$) and at the reference velocity points v_j.

However, the maximum difference amounts to only 3.5 °C. Therefore, there must to be other factors which have an influence on the temperature profiles of the tyre treads and of the transmission oil. One different factor is the different cooling air stream in the both test benches. At the flat belt dynamometer there is only a cooling fan with an area of about 0.3 m². This fan is located at a distance of about 30 cm in front of the vehicle. On the other hand, in the AEROLAB wind tunnel, the vehicle stands in front of the wind tunnel nozzle with an area of 14 m². Therefore, it can be assumed that the cooling effect in the AEROLAB wind tunnel is higher than at the flat belt dynamometer. Furthermore, using the flat belt dynamometer, the vehicle stands on four small WDUs. In contrast, the AEROLAB wind tunnel has a single-belt-rolling-road system, which is assumed to be essential for the air flow under the vehicle underbody and can be part of further cooling effects as compared to the flat belt dynamometer [60].

Difference due to the vehicle position

The influence of the vehicle position at the flat belt dynamometer on the resulting road load has already been investigated in subsection 4.1.5. For the measurements using the GTR No. 15 wind tunnel method or the wind tunnel method extended, the vehicle front and the vehicle rear are centered at the flat belts. However, in the AEROLAB wind tunnel the vehicle is only fixed at the front wheel hubs (see Figure

3.7). Therefore, only the front of the vehicle can be centered. It is assumed, that the rotation angle β of the vehicle in the AEROLAB wind tunnel corresponds to the rotation angle of $+0.15°$ at the flat belt dynamometer, where also only the front of the vehicle is centered. The resulting rolling resistance considering this different vehicle position is on average about $2.2\,\text{N}$ lower in the investigated velocity range (see Figure 4.16).

Effect of the differences in rolling resistance and drivetrain losses
In Figure 4.51 the road load determined with the AEROLAB method, but without the following influences, referenced to the coastdown method are plotted over the reference velocity points v_j:

- Belt surface (grey dashed line with open circles)

- Force measurement system (light grey dashed line with open circles)

- Vehicle position (dark grey dashed line with open circles)

- Lower transmission oil temperatures due to better cooling effects (light grey dashed line with dots)

- Vehicle fixation system (grey dashed line with dots)

- Lower tyre tread temperatures (dark grey dashed line with dots)

The curves for the coastdown method (black line with squares), the wind tunnel method extended (dark grey line with triangles) and the AEROLAB method (grey line with dots) are the same as those shown in Figure 4.29.

As discussed in the previous subsections, the three influencing factors, force measurement system, belt surface and vehicle position, would result in a lower road load as compared to the wind tunnel method due to GTR No. 15 or the wind tunnel method extended. If these effects are subtracted from the results determined using the AEROLAB method, it can be seen that the road load is increased by a maximum of 1.6 percentage points at $20\,^{\text{km}}/_{\text{h}}$, if the lower rolling resistance on the steel belt in the AEROLAB wind tunnel is compensated. There is almost no influence of the force measurement system of the AEROLAB method (integrated load cells in the vehicle fixation system) and of the vehicle position in the velocity range from $90\,^{\text{km}}/_{\text{h}}$ to $130\,^{\text{km}}/_{\text{h}}$. At $20\,^{\text{km}}/_{\text{h}}$ for both factors a maximum deviation of about 1.2 percentage points is visible. On the other hand, the lower transmission oil temperature and tyre tread temperatures due to the $3\,^{\circ}\text{C}$ lower ambient temperature and the better cooling effects in the AEROLAB wind tunnel compared to the flat belt dynamometer, as well as the vehicle fixation system at the front wheel hubs, lead to higher measured resistance forces in the AEROLAB wind tunnel. If these effects are subtracted from the AEROLAB method, it can be seen that the lower transmission oil temperature and the vehicle fixation system almost have the same effect on the resulting road load. At the $30\,^{\text{km}}/_{\text{h}}$ point, the road load is decreased by about 1.6 percentage points for both influencing factors. Nevertheless, the higher rolling resistance has the strongest

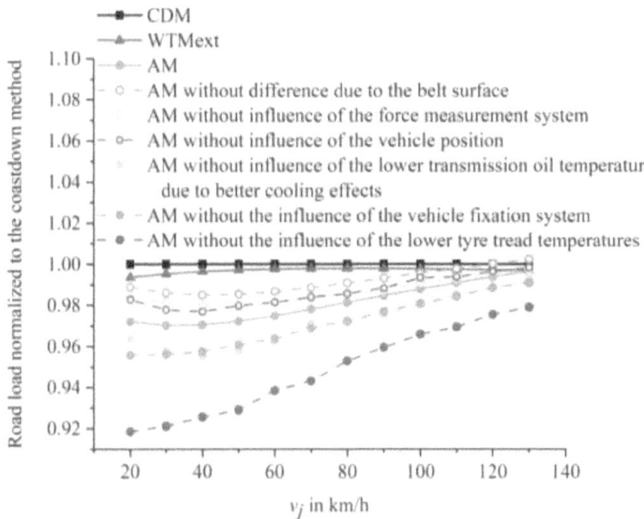

Figure 4.51: Differences in the determined road load of a vehicle between the coastdown method (black line with squares), the wind tunnel method extended (dark grey line with triangles), the AEROLAB method (grey line with dots) and the AEROLAB method without the differences in rolling resistance and drivetrain losses (dashed lines), referenced to the coastdown method and plotted over the reference velocity points v_j.

influence on the determined road load, due to the significant lower tyre tread temperatures in the AEROLAB wind tunnel as compared to the flat belt dynamometer. The maximum difference with about 8 percentage points occurs at the 20 km/h point.

Summary

In the following, the effect of the previously discussed influencing factors are evaluated with the aid of the difference in cycle energy demand (ϵ) using the WLTC and the ARTEMIS European driving cycle. In Figure 4.52, the effect of each prior discussed influencing factor on the difference in cycle energy demand ϵ is plotted as a bar chart. The differences in the aerodynamic drag are illustrated with dark grey bars, the differences in the test procedure with grey bars and the differences in the rolling resistance and drivetrain losses with light grey bars.

It is pointed out that all influencing factors concerning the aerodynamic drag have a significant impact on the cycle energy demand. By contrast, the separately determined residual brake forces, the increased stabilization time, as well as the force measurement

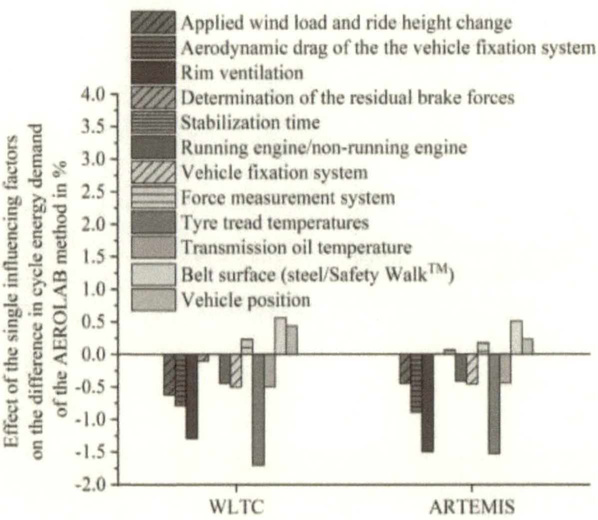

Figure 4.52: Effect of the influencing factors concerning the aerodynamic drag (dark grey bars), the test procedure (grey bars) and the rolling resistance and drivetrain losses (light grey bars) on the difference in cycle energy demand of the AEROLAB method.

system have almost no effect on the difference in energy demand. Furthermore, it can be concluded that the highest impact on the road load measured with the AEROLAB method is found for the included wheel ventilation resistance and the higher rolling resistance due to the lower tyre tread temperatures in the AEROLAB wind tunnel. Using the AERTMIS European driving cycle, these two factors have almost the same impact, of about 1.5 percentage points.

In the next step, all these influencing factors are subtracted together from the road load determined using the AEROLAB method. This means that the road load is corrected nearly to the test conditions, which exist using the wind tunnel method according to GTR No. 15 with Safety Walk™ as a rolling surface. The resulting difference in cycle energy demand using the WLTC now amounts to -5.8 % and using the ARTEMIS European driving cycle it amounts to -5.6 %. Finally, this result is now additionally corrected, taking into account that the rolling resistance for different wheel loads on asphalt is 20 % to 30 % higher than the rolling resistance on a Safety Walk™ surface [38]. For the correction, the correction factor $K_{\text{ext}} = 1.34$ is used, which was determined for the wind tunnel method extended (see subsection 4.2.1). Due to this additional correction, the difference in cycle energy demand between the to asphalt conditions corrected AEROLAB method and the coastdown method results in a value of 0.6 % based on the WLTC and 0.1 % based on the

ARTEMIS European driving cycle.

Therefore, it can be concluded that there are several influencing factors, as for example the additional wheel ventilation resistance and the higher rolling resistance due to lower tyre tread temperatures, which affect the road load of a vehicle determined using the AEROLAB method in such a positive way, that it is consistent with the road load determined using the coastdown method on a proving ground. However, if the influencing factors due to these differences are subtracted from the road load according to the AEROLAB method, there is also the need of an additional correction factor K_{ext}, which is defined for the wind tunnel method extended in subsection 4.2.1, to reach the same rolling resistance level as the one on a real road.

Additionally, it has to be clarified that the correction factor K_{ext} only considers the correction of the rolling resistance, due to different surface characteristics of the belts in the test bench and of the real road on the test track. However, in Figure 4.52, it is shown that especially the ride height change, in combination with the wind velocity and the wheel ventilation resistance, have a significant influence on the difference in the cycle energy demand. In addition, these effects exist also during the coastdown runs on the test track and affect the resulting road load, accordingly. However, these factors are not taken into account, neither in the wind tunnel method according to GTR No. 15 nor in the wind tunnel method extended.

Furthermore, it is pointed out that it is not sufficient to consider only the ambient temperature for the correction of the rolling resistance to reference conditions (see Equation 2.45). In Figure 4.47 it is shown that there is a difference in the tyre tread temperatures of about 13 °C, although the difference in ambient temperature in the wind tunnel and at the flat belt dynamometer only amounts to 3 °C. These results are consistent with the statements made in [19], where it was shown that the absolute value of the rolling resistance coefficient directly depends on the tyre tread temperature, and these in turn on the road surface temperatures.

4.3.2 Verification of the AEROLAB method

Finally, the AEROLAB method is verified with the BMW G30 518d and the MINI F60 Countryman. These two vehicles were already used for the verification of the wind tunnel method extended (see subsection 4.2.2). In Figure 4.53, the road load for the BMW G30 518d (light grey lines with triangles) and the F60 Countryman (grey lines with dots) determined with the AEROLAB method (open symbols) and with the wind tunnel method extended (filled symbols) normalized with the corresponding road load determined with the coastdown method, are plotted over the reference velocity points v_j. The results determined with the wind tunnel method extended (WTMext) were already presented in Figure C.1.

It can be seen that the AEROLAB method results in an underprediction of the road load for the G30 518d and in an overprediction of the F60 Countryman. The maximum deviation from the coastdown method for both vehicles occurs at the 20 $^{\text{km}}/_{\text{h}}$ reference velocity point.

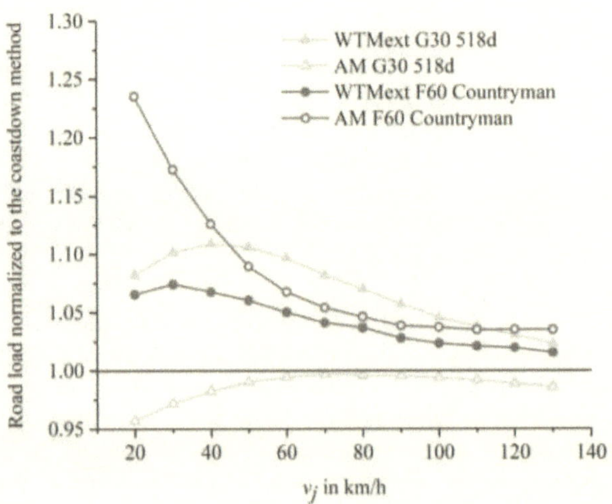

Figure 4.53: Road load for the BMW G30 518d (light grey lines with triangles) and the F60 Countryman (grey lines with dots) determined with the AEROLAB method (open symbols) and with the wind tunnel method extended (filled symbols) normalized with the corresponding road load determined with the coastdown method plotted over the reference velocity points v_j.

The differences amount to about 24 % for the MINI F60 Countryman and nearly 5 % for the BMW G30 518d.

In addition, it is shown that the deviation between the results according to the wind tunnel method extended increases with decreasing velocity for both vehicles. The same characteristic is also found for the BMW F46 216i (see Figure 4.29).

The increase of the road load of the F60 Countryman with decreasing velocity, measured using the AEROLAB method compared to the coastdown method, is considered more precisely in the following. In Figure 4.54, the residual brake forces $F_{j,\mathrm{Brake}}$ of the F60 Countryman determined using the test procedure described by Equation 3.7 at the flat belt dynamometer are illustrated over the reference velocity points. It can be seen that the residual brake forces are also increasing with decreasing velocity. Although it was ensured that the brake pads do not drag on the brake disks prior to each measurement in the AEROLAB wind tunnel, it is assumed that the results of the AEROLAB wind tunnel already contain residual brake forces. If these residual brake forces determined using the flat belt dynamometer are subtracted from the road load determined using the AEROLAB method, the maximum deviation is reduced from about 24 % to about 16 % at the $20\,\mathrm{km/h}$ reference velocity point (see Figure D.4 in the appendix).

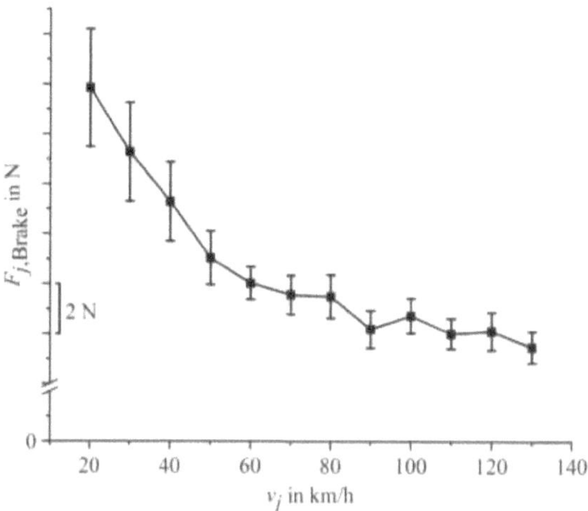

Figure 4.54: Residual brake forces $F_{j,\text{Brake}}$ for the F60 Countryman plotted over the reference velocity points v_j.

According to the detailed analysis of the factors which influence the absolute road load values of the F46 216i, it is pointed out that mainly the lower tyre tread temperatures (see in Figure 4.51, dark grey dashed line with dots), the lower oil transmission temperatures due to both the better cooling effect in the AEROLAB wind tunnel (see in Figure 4.51, light grey dashed line with dots) and the non-running engine (see in Figure 4.39, grey dashed line with open circles) have an increasing impact with decreasing velocity on the road load determined with the AEROLAB method. Additionally, it is shown that, although the ambient temperature between the AEROLAB wind tunnel and the flat belt dynamometer amounts to only 3 °C, the differences in the tyre tread temperatures and the oil transmission temperatures measured in both test benches are significantly higher (see Figure 4.47). On this account, the tyre temperatures and oil transmission temperatures at the end of the warm-up phase are also compared for the verification vehicles F60 Countryman and G30 518d. However, in this case there are no infrared sensors mounted in the wheel housings for the determination of the tyre tread temperatures. Instead, the Tire Pressure Monitoring Systems (TPMS) are used (see subsection 3.3.2). Consequently, the correlation between the tyre tread temperature measured with the infrared sensors mounted in the wheel housings and the tyre air temperature measured with the TPMS is investigated using the test vehicle F46 216i, initially. For this purpose, the tyre temperatures at the end of the warm-up phase of some coastdown runs are measured with both systems (infrared sensors in the wheel housings and TPMS). Furthermore, the coastdown runs are executed at different ambient temperatures. In Figure 4.55, both the tyre air temperatures measured with the TPMS (open symbols) and the tyre tread temperatures measured with the infrared sensors (filled symbols) for the tyres front left FL (black symbols) and rear left RL (grey symbols) are plotted over the ambient temperature for the test vehicle F46 216i.

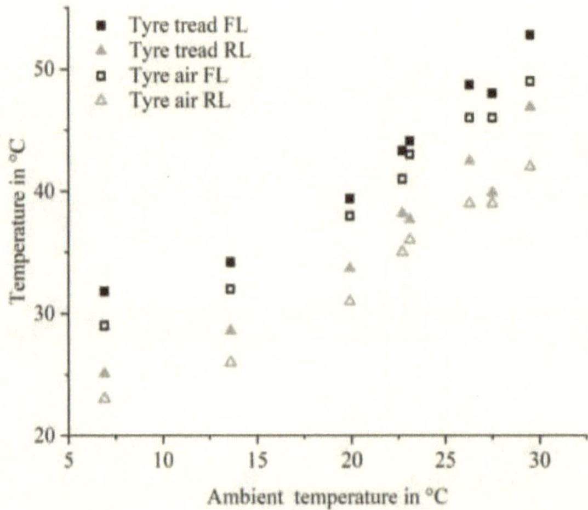

Figure 4.55: Tyre air temperature (open symbols) measured with the Tire Pressure Monitoring System and the tyre tread temperatures (filled symbols) measured with infrared sensors installaed in the wheel housings for the tyres front left FL (black symbols) and rear left RL (grey symbols) plotted over the ambient temperature for the test vehicle F46 216i.

It can be seen that the absolute values are different, while the curve characteristics are the same. Furthermore, it is pointed out that with increasing ambient temperature, both the tyre tread and the tyre air temperatures are also increasing. The same illustration, but for the tyres front right (FR) and rear right (RR), is given in Figure D.5 in the appendix. Following, it can be stated that with the aid of the TPMS, the same qualitative statement about the tyre temperature conditions at the end of the warm-up phase can be obtained as with the infrared sensors. Considering that, in Figure 4.56 the tyre air temperature for all four tyres and the transmission oil temperature, or rather the rear wheel drive temperatures, for the F60 Countryman (on the left side) and for the G30 518d (on the right side) in the AEROLAB wind tunnel (light grey bars) and at the flat belt dynamometer (grey bars), are illustrated at the end of the warm-up phase.
In this illustration the same behaviour as for the F46 216i can be observed. Although the ambient temperatures differ only by $3\,°\mathrm{C}$ between both test benches, the tyre temperatures and the transmission/rear wheel drive oil temperatures are $6\,°\mathrm{C}$ to $13\,°\mathrm{C}$ lower in the AEROLAB wind tunnel than at the flat belt dynamometer. Due to these results, it can

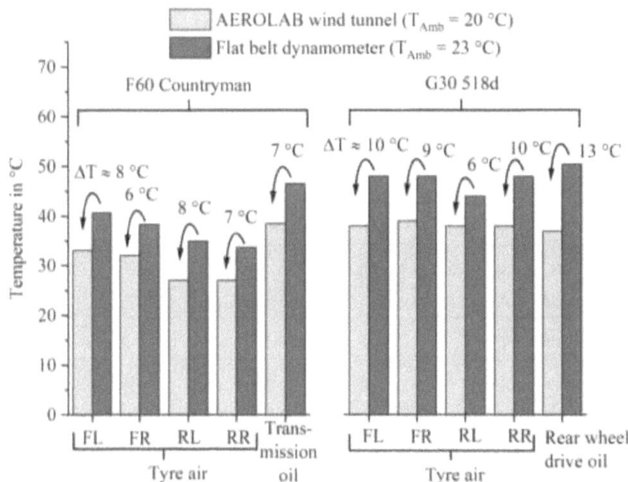

Figure 4.56: Tyre air temperature and transmission oil temperature of rear wheel drive oil temperature for the test vehicle F46 216i (left) and G30 530d (right) measured in the AEROLAB wind tunnel at 20 °C (light grey bars) and at the flat belt dynamometer at 23 °C (grey bars).

be again assumed that the drivetrain losses and the rolling resistance determined with the AEROLAB method are higher than when determined with the wind tunnel method according to GTR No. 15 or with the wind tunnel method extended.

Finally, the AEROLAB method is again evaluated using the differences in the cycle energy demand ϵ based on the WLTC and the ARTEMIS European driving cycle (see Equation 3.9) for both vehicles. The results are given in Table 4.7. The differences in cycle energy demand between the wind tunnel method extended and the coastdown method $\epsilon_{\mathrm{WTMext}}$ are the same as those already stated in Table 4.5.

It can be seen that the differences in cycle energy demand for the F60 Countryman are lower for both driving cycles using the wind tunnel method extended than for the AEROLAB method. On the other hand, the differences in cycle energy demand for the G30 518d are reduced using the road load determined with the AEROLAB method. The maximum difference amounts to only -0.7 %. Furthermore, all differences in cycle energy demand lie again between ±5 %, which is also an acceptance criterion according to GTR No. 15 [10].

Table 4.7: Differences in energy demand between the wind tunnel method extended with $K_{\text{ext}} = 1.34$ and the coastdown method ϵ_{WTMext} and between the AEROLAB method and the coastdown method ϵ_{AM} for the vehicles F60 Countryman and G30 518d

Difference in cycle energy demand in %	WLTC	ARTEMIS
F60 Countryman		
ϵ_{WTMext}	1.9	1.6
ϵ_{AM}	3.2	2.9
G30 518d		
ϵ_{WTMext}	3.2	2.6
ϵ_{AM}	-0.6	-0.7

4.4 Error calculation

In the following, the quality of the different methods for road load determination is discussed further with the aid of an error calculation. The error calculation is based on GUM, which was already introduced in section 2.6.

4.4.1 Coastdown method

One of the principal tasks of an error calculation according to GUM is to find a mathematical relationship between the measurand Y and the input quantities X_1, X_2, ... X_m (see Equation 2.76). For the road load determination using the coastdown method, the road load is determined based on the measured vehicle masses at the beginning and at the end of the test, as well as on the velocity and on the elapsed time during the test:

$$F_j = (m_{\text{av}} + m_{\text{r}}) \cdot \frac{\Delta v}{\Delta t_j} \tag{4.7}$$

where:

F_j is the uncorrected road load of the vehicle at the reference velocity point v_j in N (see Equation 2.36);

m_{av} is the arithmetic average of the test vehicle masses at the beginning and end of the coastdown test procedure in kg (see Equation 2.38);

m_{r} is the the equivalent effective mass of rotating components (for more information see in [10]) in kg;

Δv is the incremental step between two reference velocity points ($= 10\,{}^{\text{km}}/\text{h}$) in ${}^{\text{m}}/\text{s}$ (see Equation 2.39);

Δt_j is the harmonic average of the alternate coastdown time measurements in direction A and B at the reference velocity point v_j in s (see Equation 2.42).

Afterwards, the road load coefficients f_0, f_1 and f_2 (see Equation 2.2) are calculated with a polynomial regression of second order and then the correction to reference conditions is invoked (compare Equation 2.45):

$$F_{j,\text{CDM}} = ((f_0 - w_1 - K_1) + f_1 \cdot v_j) \cdot \left(1 + K_0 \cdot \left(\overline{T}_{\text{Amb}} - T_0\right)\right) + K_2 \cdot f_2 \cdot v_j^2 \qquad (4.8)$$

where:

$F_{j,\text{CDM}}$	is the road load of the vehicle corrected to reference conditions at the vehicle velocity v according to the coastdown method (CDM) in N (see Equation 2.45);
f_0	is the constant term of the road load coefficients in N;
f_1	is the coefficient of the first order term of the road load coefficients in $\text{N}/\text{(m/s)}$;
f_2	is the coefficient of the second order term of the road load coefficients in $\text{N}/\text{(m/s)}^2$;
K_0	is the correction factor for rolling resistance (and drivetrain losses) in K^{-1} (see Equation 2.46);
K_1	is the test mass correction factor in N (see Equation 2.47);
K_2	is the correction factor for air resistance in N (see Equation 2.48);
$\overline{T}_{\text{Amb}}$	is the arithmetic average ambient atmospheric temperature in K;
T_0	is the reference atmospheric temperature of $20\,°\text{C}$ in K;
v_j	is the reference velocity point in m/s;
w_1	is the wind resistance due to wind of opposite directions alongside the road of the test track during the measurement procedure in N.

Due to this polynomial regression, it is not possible to summarize the Equations 4.7 and 4.8 to one single mathematical relationship according to Equation 2.76. Therefore, the error calculation has to be divided into two parts. The first error calculation can be made based on Equation 4.7 (before polynomial regression) and afterwards based on Equation 4.8 (after polynomial regression).

Before polynomial regression

However, the error calculation 'Before polynomial regression' is left out in this study as the result of the error calculation of the uncorrected road load can not be compared to the results of the other methods which concerne to reference conditions corrected road loads. Furthermore, the uncertainties of the velocity and time determination of the GPS (Global Positioning System) device are included in the imprecisions of the uncorrected road load coefficients determined in the following subsection. The uncertainty due to the vehicle weighing system is also part of the error calculation 'After polynomial regression'.

For the sake of completeness, the technical data of the GPS device for the velocity and time determination are stated in the following: The vehicle velocity and the coastdown time are recorded via a GPS device, which collects the data with a frequency of $10\,\text{Hz}$[10]. Additionally, the velocity of the vehicle is determined with an accuracy of $\pm0.1\,\text{m/s}$ ($\pm0.36\,\text{km/h}$)[10].

The uncertainty of the vehicle weighing system is stated in the next subsection.

After polynomial regression

Initially, all influencing quantities δx_m are inserted into Equation 4.8 and into the corresponding correction terms defined in subsection 2.3.1 (see Equations 2.49 to 2.54) to describe the relationship between the measurand Y and the input quantities X_m (compare Equation 2.76). Afterwards, the influencing quantities are discussed.

$$F_{j,\text{CDM}} = \delta F_{\text{corr}} +$$
$$\left(\left(\left(f_0 + \delta f_0 - w_1 - K_1\right) + \left(f_1 + \delta f_1\right) \cdot v_j\right) \cdot \left(1 + K_0 \cdot \left(\overline{T}_{\text{Amb}} + \delta T_{\text{Amb}} - T_0\right)\right)\right.$$
$$\left. + K_2 \cdot \left(f_2 + \delta f_2\right) \cdot v_j^2\right) \quad (4.9)$$

with

$$K_0 = 8.6 \cdot 10^{-3}\,\text{K}^{-1} \quad (4.10)$$

$$K_1 = \left(f_0 + \delta f_0\right) \cdot \left(1 - \frac{m_{\text{test}}}{m_{\text{av}} + 2 \cdot \delta m}\right) \quad (4.11)$$

$$K_2 = \frac{\overline{T}_{\text{Amb}} + \delta T_{\text{Amb}}}{293\,\text{K}} \cdot \frac{100\,\text{kPa}}{\overline{p}_{\text{Amb}} + \delta p_{\text{Amb}}} \quad (4.12)$$

and

$$w_1 = \left(f_2 + \delta f_2\right) \cdot \left(v_{\text{Wind}} + \delta v_{\text{Wind}}\right)^2 \quad (4.13)$$

where:

$\delta f_0,\ \delta f_1,\ \delta f_2$ are the influencing quantities of the road load coefficients of the uncorrected road load F_j in N, $\text{N}/\text{(m/s)}$ and $\text{N}/\text{(m/s)}^2$;

δT_{Amb} is the influencing quantity of the measured ambient temperature T_{Amb} in K;

δm is the influencing quantity of the vehicle weighing system to determine the vehicle weight before and after the test procedure in kg;

δp_{Amb} is the influencing quantity of the measured ambient pressure p_{Amb} in kPa;

δv_{Wind} is the influencing quantity of the measured wind velocity v_{Wind} in m/s;

δF_{corr} is the influencing quantity of the corrected road load determined at different days in N.

Influencing quantity of the measured ambient conditions δT_{Amb}, δp_{Amb} and δv_{Wind}

The stationary anemometer, located next to the test track, measures the ambient temperature T_{Amb}, the ambient pressure p_{Amb} and the wind velocity v_{Wind}. The expanded uncertainty for the ambient temperature δT_{Amb} is $\pm 0.03\,\text{K}$[23], for the ambient pressure $\delta p_{\text{Amb}}\ \pm 7\,\text{Pa}$[24] and for the wind velocity $\delta v_{\text{Wind}}\ \pm 8\,\text{mm/s}$[25]. The coverage factor k_p for these uncertainties is 2, respectively. All uncertainties are taken from calibration certificates and are therefore of type B.

Influencing quantity of the vehicle weighing system δm

For the correction of the road load to reference conditions, the vehicle weight has to be determined before and after the test procedure. Therefore, the uncertainty does not change and has a value of $\pm 2\,\text{kg}$[26]. In Equation 4.11, this uncertainty is considered twice, since the vehicle mass has to be measured at the beginning and at the end of the measurement procedure. This uncertainty, taken from a calibration certificate, is of type B.

Influencing quantity of the uncorrected road load coefficients δf_0, δf_1 and δf_2

For the investigated coastdown measurement, six single coastdown runs are executed, whereby each coastdown run consists of split runs in both directions A and B, as shown in Figure 4.57 in Step 1. To estimate the imprecisions of the uncorrected road load coefficients, for each coastdown run the uncorrected road load F_j as well as the average and its standard deviation over the six coastdown runs (Step 2) are determined. In the next step (Step 3), the road load coefficients for the averaged road load curve and the averaged road load curves plus and minus its standard deviation are calculated. Afterwards (Step 4), the maximal differences $\Delta f_{0,\text{max}}$, $\Delta f_{1,\text{max}}$ and $\Delta f_{2,\text{max}}$ between the corresponding road load coefficients are determined. Finally, half of the maximal difference is defined as the imprecision of the corresponding road load coefficient.

Therefore, the uncertainty for the road load coefficient f_0 amounts $\pm 3.22\,\text{N}$, for f_1 $\pm 0.576\,\text{N/(m/s)}$ and for f_2 $\pm 0.03253\,\text{N/(m/s)}^2$. Due to the rectangular PDFs (see Figure 4.57), all three uncertainties are also a type B evaluation. The above defined uncertainties of the coefficient include the following factors:

- Uncertainty due to the time and velocity measurements with the GPS device (compare the previous subsection 'Before polynomial regression')

- Continuous weight loss during the entire measurement procedure: The correction only considers the average of the vehicle mass at the beginning and at the end of the test.

---

[23]Vaisala - Measurement Standards Laboratory - Accredited Calibration Laboratory. Certificate of calibration: Humidity and Temperature Probe, 2019.

[24]Vaisala OYI. Calibration certificate: BAO-1QML-AH, 2019.

[25]Deutsche WindGuard- Wind Tunnel Serlives GmbH. Calibration certificate: 2D Sonic Anemometer, 2019.

[26]WTM Wägetechnik. Kalibrierschein: Bodenwaage, 2017.

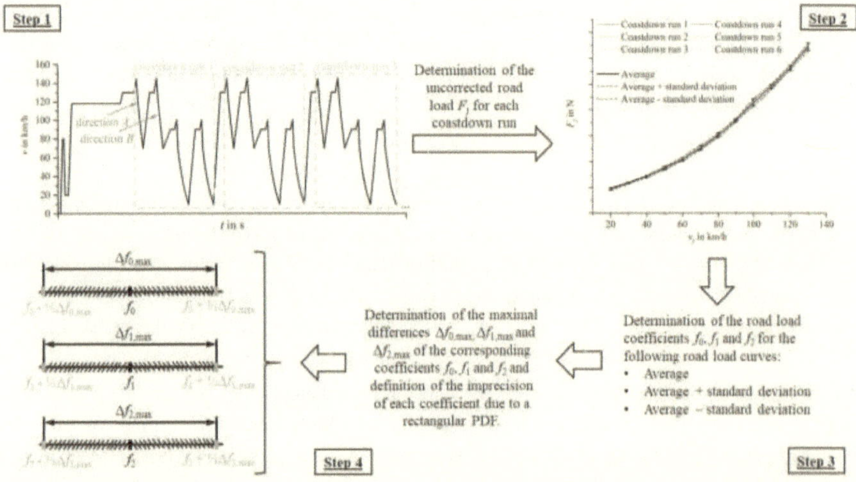

Figure 4.57: Determination of the imprecisions of the road load coefficients f_0, f_1 and f_2 for the error calculation of the coastdown method (after polynomial regression).

- Fluctuations of air temperature, air pressure as well as wind velocity and direction during the entire measurement procedure: The corrections to reference conditions of the road load are made on basis of the average air temperature, average air pressure and average wind velocity for each coastdown run. The fluctuations during the run are not considered.

- Little steering maneuvers, which are always necessary to drive on a straight line

- Difference in road surface temperatures (see Figure 2.14 and in [19]) and road surface fluctuations, which have a direct influence on the resulting rolling resistance (see Figure 2.14 and in [19])

Furthermore, the determined road load coefficients are not independent quantities. Therefore, the correlation term has to be taken into account (see Equation 2.90). In Figure 4.58 the correlations between the road road load coefficients f_0, f_1 and f_2 are illustrated. The corresponding correlation coefficients $r\left(x_m, x_l\right)$ are determined using polynomial trendlines:

- (a): Correlation between f_0 and f_1 with $r\left(f_0, f_1\right) = -0.9799$

- (b): Correlation between f_0 and f_2 with $r\left(f_0, f_2\right) = 0.9312$

- (c): Correlation between f_1 and f_2 with $r\left(f_1, f_2\right) = -0.9971$

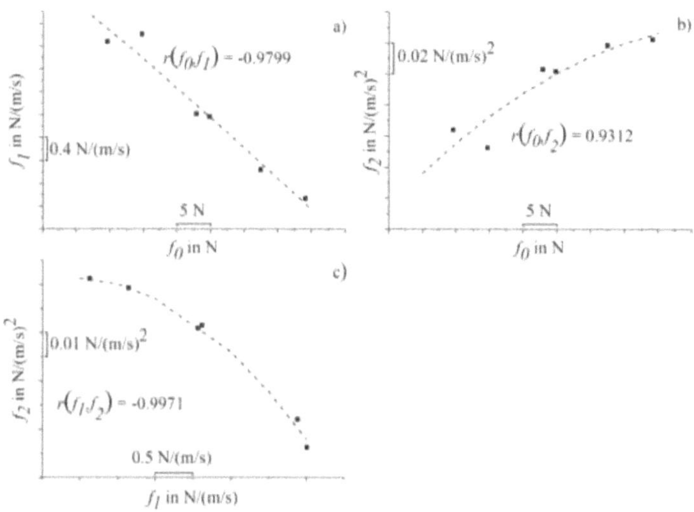

Figure 4.58: Road load coefficients f_1 over f_0 (a), f_2 over f_0 (b) and f_2 over f_1 (c) with their corresponding polynomial trendlines and their correlation coefficients $r\left(x_m, x_l\right)$.

Influencing quantity of the corrected road load δF_{corr}

Up to now, only the uncertainties of one single coastdown measurement are considered. Therefore, the corrected road loads of in total 10 coastdown runs executed at two different days are determined. Afterwards, the standard deviation over these road load curves must be determined. This standard deviation σ depends on the velocity and is stated in Table 4.8.

Table 4.8: Standard deviation $\sigma\,(F_{\mathrm{corr}})$ of the corrected road load curves of ten single coastdown runs executed at two different days

v_j in km/h	20	30	40	50	60	70	80	90	100	110	120	130
$\sigma\,(F_{\mathrm{corr}})$ in N	4.8	3.6	3.1	3.1	3.2	3.1	3.0	3.0	3.5	4.7	6.7	9.2

Finally, the corresponding standard uncertainty is determined using the standard deviation and is therefore of type A.

The resulting expanded uncertainty U_y with $k_p = 2$ for each reference velocity point v_j in percent and referenced to the road load determined according to the coastdown method are listed in Table 4.9. It can be seen that the uncertainty is mostly increasing with increasing velocity. The maximum uncertainty with a value of 3.8 % occurs at the reference velocity point of 130 km/h.

Table 4.9: Expanded uncertainty U_y with $k_p = 2$ for the coastdown method (after polynomial regression) for each reference velocity point v_j in percent and referenced to the road load determined using the coastdown method

v_j in km/h	20	30	40	50	60	70	80	90	100	110	120	130
U_y in %	1.6	1.0	0.7	0.7	1.0	1.3	1.8	2.2	2.6	3.1	3.5	3.8

Furthermore, to identify the impact of the single uncertainties on the result for the expanded uncertainty U_y listed in Table 4.9, the values of the modified components of the combined standard uncertainty $u^*_{x_m,y}$ (calculation based on Equation 2.91) are plotted over the reference velocity point v_j in Figure 4.59. The corresponding equations for the sensitivity coefficients c_m are given in the appendix (see Equations E.1 to E.8). However, the real values of the components of the combined standard uncertainty $u_{x_m,y}$ are not declared. Instead, the original values are multiplied with a constant factor τ, whereby its value is unknown to the reader. The modified components of the combined standard uncertainty $u^*_{x_m,y}$ are defined as:

$$u^*_{x_m,y} = u_{x_m,y} \cdot \tau \tag{4.14}$$

Furthermore, the modified components of the combined standard uncertainty $u^*_{x_m,y}$ for the input quantities m_1 and m_2 are added together, as their single absolute values are the same:

$$u^*_{m,F_{j,CDM}} = u^*_{m_1,F_{j,CDM}} + u^*_{m_2,F_{j,CDM}} \qquad (4.15)$$

where:

$u^*_{x_m,y}$ are the modified components of the combined standard uncertainty $u^*_{x_m,y}$ in N;

$u_{x_m,y}$ are the components of the combined standard uncertainty $u^*_{x_m,y}$ in N;

τ is a constant factor, whereby its value is unkown to the reader;

$u^*_{m,F_{j,CDM}}$ is the modified component of the combined standard uncertainty for the determination of the vehicle weight before and after the measurements using the coastdown method (after polynomial regression) in N;

$u^*_{m_1,F_{j,CDM}}$ is the modified component of the combined standard uncertainty for the determination of the vehicle weight m_1 before the measurement using the coastdown method (after polynomial regression) in N;

$u^*_{m_2,F_{j,CDM}}$ is the modified component of the combined standard uncertainty for the determination of the vehicle weight m_2 after the measurement using the coastdown method (after polynomial regression) in N.

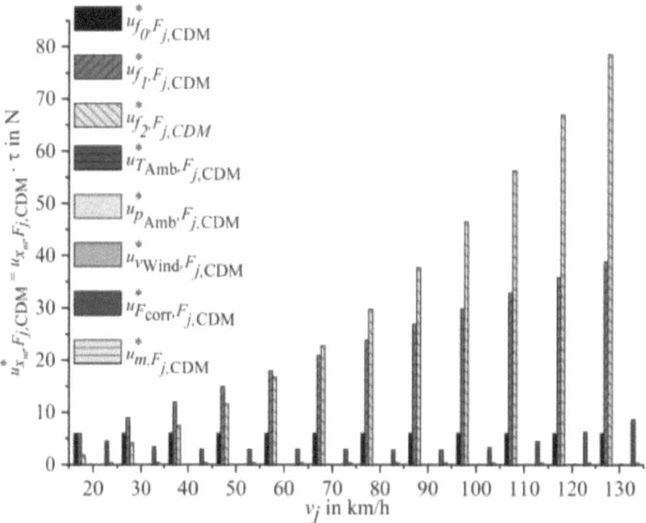

Figure 4.59: The modified components of the combined standard uncertainties $u^*_{x_m,F_{j,CDM}}$ for the influencing quantities δx_m of the coastdown method (after polynomial regression).

It can be seen that the uncertainties due to the uncorrected road load coefficients f_1 and f_2 have the greatest impact on the expanded uncertainty up to a value of the modified

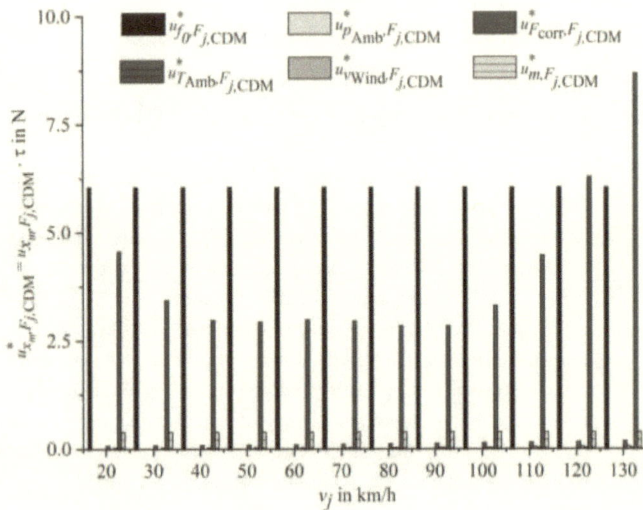

Figure 4.60: Enlarged section of Figure 4.59 without $u^*_{f_1,F_{j,\mathrm{CDM}}}$ and $u^*_{f_2,F_{j,\mathrm{CDM}}}$: The modified components of the combined standard uncertainties $u^*_{x_m,F_{j,\mathrm{CDM}}}$ for the influencing quantities δx_m of the coastdown method (after polynomial regression).

component of the combined uncertainty $u^*_{f_2,F_{j,\mathrm{CDM}}}$ of about 78.6 N[27] for the f_2 coefficient at the reference velocity point of 130 km/h. This can be explained by the simple and the quadratic velocity dependency of the factors f_1 and f_2. The uncertainty for the first road load coefficient f_0 is independent of the velocity. Thus, $u^*_{f_0,F_{j,\mathrm{CDM}}}$ has a constant value of about 6 N[27]. The uncertainty due to the corrected road load curves on two different days $u^*_{F_{\mathrm{corr}},F_{j,\mathrm{CDM}}}$ has a value of less than 10 N[27]. The other components $u^*_{T_{\mathrm{Amb}},F_{j,\mathrm{CDM}}}$, $u^*_{p_{\mathrm{Amb}},F_{j,\mathrm{CDM}}}$, $u^*_{v_{\mathrm{Wind}},F_{j,\mathrm{CDM}}}$ and $u^*_{m,F_{j,\mathrm{CDM}}}$ have even a lower value, as pointed out in Figure 4.60, in which an enlarged section of Figure 4.59 without $u^*_{f_1,F_{j,\mathrm{CDM}}}$ and $u^*_{f_2,F_{j,\mathrm{CDM}}}$ is illustrated. Thus, their impact on the resulting expanded uncertainty is negligible. For further information, the values of all components are listed in Table E.1 in the appendix.

---

[27]This value is not the real value for the component of the combined standard uncertainty. Instead, it is multiplied with the factor τ (see Equation 4.14).

4.4.2 Wind tunnel method extended

The road load determination using the wind tunnel method extended is divided into two parts: the determination of the aerodynamic drag in the BMW Group wind tunnel and the determination of both the drivetrain losses and the rolling resistance at the flat belt dynamometer (see Equation 3.5). Therefore, the error calculation is also divided into these two parts.

Aerodynamic drag

The aerodynamic drag is calculated using Equation 3.4, whereby the aerodynamic drag coefficient c_w is determined using the Equations 3.1 to 3.3. In the following, the influencing quantities are inserted into these equations to describe the relationship between the measurand Y and the input quantities X_m (compare Equation 2.76). Afterwards, the influencing quantities are discussed.

$$c_\mathrm{w} = \frac{2 \cdot (F_{x,\mathrm{Wind}} + \delta F_{x,\mathrm{acc}} + \delta F_{x,\mathrm{res}} + \delta F_{x,\mathrm{rep}} + \delta F_{x,\mathrm{fluc}})}{\rho_\mathrm{TS} \cdot A_x \cdot U_\mathrm{TS}^2} \tag{4.16}$$

with

$$U_\mathrm{TS} = \sqrt{\frac{2 \cdot q_\mathrm{TS}}{\rho_\mathrm{TS}}} \tag{4.17}$$

and

$$q_\mathrm{TS} = \frac{\kappa \cdot (p_\mathrm{stat} + \delta p_\mathrm{s} + \delta p_\mathrm{s,fluc})}{\kappa - 1} \cdot \left[\left(\frac{\Delta p + \delta \Delta p + \delta \Delta p_\mathrm{fluc}}{p_\mathrm{stat} + \delta p_\mathrm{s} + \delta p_\mathrm{s,fluc}} + 1 \right)^{\frac{\kappa - 1}{\kappa}} - 1 \right] \tag{4.18}$$

where:

c_w	is the aerodynamic drag coefficient (see Equation 3.1);
$F_{x,\mathrm{Wind}}$	is the measured force of the aerodynamic drag in x direction in N;
$\delta F_{x,\mathrm{acc}}$	is the influencing quantity due to the accuracy of the force measurement of the wind tunnel balance in N;
$\delta F_{x,\mathrm{res}}$	is the influencing quantity due to the resolution of the force measurement of the wind tunnel balance in N;
$\delta F_{x,\mathrm{rep}}$	is the influencing quantity due to the repeatability of the force measurement over six measurements in N;
$\delta F_{x,\mathrm{fluc}}$	is the influencing quantity due to force measurement fluctuations in the wind tunnel in N;
ρ_TS	is the density of the air in the test section in $\mathrm{kg/m^3}$;
A_x	is the frontal area of the vehicle in $\mathrm{m^2}$;
U_TS	is the air velocity in the test section of the wind tunnel in $\mathrm{m/s}$ (see Equation 3.2);
q_TS	is the dynamic pressure in the test section of the wind tunnel in Pa (see Equation 3.3);

κ	is the ratio of the specific heats for air and has a value of 1.4;
p_{stat}	is the absolute static pressure measured in the plenum of the wind tunnel in Pa;
δp_{s}	is the influencing quantity due to the accuracy of the static pressure measurement in Pa;
$\delta p_{\mathrm{s,fluc}}$	is the influencing quantity due to the fluctuation of the static pressure measurement in Pa;
Δp	is the differential pressure between the total and static pressure port in Pa;
$\delta \Delta p$	is the influencing quantity due to the accuracy of the differential pressure measurement in Pa;
$\delta \Delta p_{\mathrm{fluc}}$	is the influencing quantity due to the fluctuation of the static pressure measurement in Pa.

Influencing quantities related to the force measurement in x direction

For the c_{w} determination, the force in x direction has to be measured with the balance in the BMW Group wind tunnel. The balance has an accuracy of $\pm 0.97\,\mathrm{N}$ $(\delta F_{x,\mathrm{acc}})$[28] and a resolution of $\pm 0.39\,\mathrm{N}$ $(\delta F_{x,\mathrm{res}})$[28]. These uncertainties are taken from calibration certificates and are therefore of type B. Additionally, fluctuations in the force measurement can be observed. These amount to a standard deviation of $\pm 6.2\,\mathrm{N}$ $(\delta F_{x,\mathrm{fluc}})$ for the test vehicle F46 216i. The measurement time is 60 seconds with a measurement frequency of 128 Hz. Reasons for these fluctuations could be small wind velocity fluctuations in the wind tunnel, possible flow separations on the vehicle and deformations of the external shell of the vehicle. Finally, the repeatability of the force measurement over six measurements is investigated. It amounts to $\pm 0.77\,\mathrm{N}$ $(\delta F_{x,\mathrm{rep}})$. These quantities are estimated with the corresponding standard deviation and are therefore of type A. However, the value of the uncertainty $\delta F_{x,\mathrm{rep}}$ is in the range of the accuracy of the balance and can be excluded from the computation, accordingly.

Influencing quantities related to the static differential pressure measurement

Both the static and the differential pressures are measured with an accuracy of $\pm 0.0012\,\mathrm{Pa}$ $(\delta p_{\mathrm{s}}$ and $\delta \Delta p)$[29]. The fluctuation during the investigated measurement amounts to $\pm 11.87\,\mathrm{Pa}$ for the static pressure $(\delta p_{\mathrm{s,fluc}})$ and $\pm 0.86\,\mathrm{Pa}$ for the differential pressure $(\delta \Delta p_{\mathrm{fluc}})$. The measurement time is 60 seconds with a measurement frequency of 128 Hz. The accuracy statements are again taken from calibration certificates and are therefore of type B. The fluctuations are evaluated due to the standard deviation during the measurement of the pressure and are therefore of type A.

---

[28]MTS. Wind Tunnel Balance Calibration Report: Summary of One-site Balance Calibration, 2019.
[29]MKS INSTRUMENTS INC. MKS Type 690A Absolute and Type 698A: Differential High Accuracy Pressure Transducers, 2009.

Rolling resistance and drivetrain losses

The measured forces with the flat belt dynamometer are corrected to reference conditions using Equations 2.53 to 2.55. The corresponding influencing quantities are inserted into these equations to describe the relationship between the measurand Y and the input quantities X_m (compare Equation 2.76):

$$F_{j,\text{Dyno}} = \left(F_{j,\text{Dyno}}^* + \delta F + \delta F_{j,\text{Dyno}} + \delta p_{\text{Tyre}} + \delta F_{\text{fluc}} - K_1 \right) \cdot$$
$$\left[1 + K_0 \{ \left(\overline{T}_{\text{Dyno}} + \delta T_{\text{Dyno}} + \delta T_{\text{Dyno,fluc}} \right) - 293\,\text{K} \} \right] \quad (4.19)$$

with

$$K_0 = 8.6 \cdot 10^{-3}\,\text{K}^{-1} \quad (4.20)$$

and

$$K_1 = \left(F_{j,\text{Dyno}}^* + \delta F + \delta F_{j,\text{Dyno}} + \delta p_{\text{Tyre}} + \delta F_{\text{fluc}} \right) \cdot \left(1 - \frac{m_{\text{test}}}{m_{\text{av}} + 2 \cdot \delta m} \right) \quad (4.21)$$

Finally, the resulting road load according to the wind tunnel method extended is based on Equation 4.22:

$$F_{j,\text{WTMext}} = \left(F_{j,\text{Dyno}} - F_{j,\text{Drive}} + \delta F_{\text{TOM}} \right) \cdot K_{\text{ext}} + F_{j,\text{Drive}} + \delta F_{\text{TOM}} + F_{j,\text{Air}} \quad (4.22)$$

where:

$F_{j,\text{Dyno}}$	is the corrected road load of the vehicle to reference conditions determined according to the wind tunnel method at the reference velocity point v_j in N (see Equation 2.53);
$F_{j,\text{Dyno}}^*$	is the uncorrected road load of the vehicle at the reference velocity point v_j determined using the wind tunnel method in N;
δF	is the influencing quantity due to the accuracy of the load cells mounted at the WDUs of the flat belt dynamometer in N;
$\delta F_{j,\text{Dyno}}$	is the influencing quantity of different road load measurements with braking phase at the flat belt dynamometer in N;
δp_{Tyre}	is the influencing quantity in the rolling resistance due to the adjustment accuracy of the tyre inflation pressure in N;
δF_{fluc}	is the influencing quantity due to fluctuations of the force measurement at the flat belt dynamometer in N;
K_1	is the test mass correction factor in N (see Equation 2.55);
K_0	is the correction factor for rolling resistance (and drivetrain losses) in K^{-1} (see Equation 2.54);
$\overline{T}_{\text{Dyno}}$	is the arithmetic average temperature at the test bench during the measurement procedure in K;

δT_{Dyno}	is the influencing quantity due to the accuracy of the temperature measurement at the flat belt dynamometer in K;
$\delta T_{\mathrm{Dyno,fluc}}$	is the influencing quantity due to fluctuation of the ambient temperature during the measurement at the flat belt dynamometer in K;
m_{test}	is the vehicle test mass defined for the road load determination in kg;
m_{av}	is the arithmetic average of the test vehicle masses at the beginning and end of the used test procedure in kg (see Equation 2.38);
δm	is the influencing quantity of the vehicle weighing system to determine the vehicle weight before and after the test procedure in kg;
$F_{j,\mathrm{WTMext}}$	is the road load of a vehicle according to the wind tunnel method extended with the additional correction factor K_{ext} in N (see Equation 3.5);
$F_{j,\mathrm{Drive}}$	are the drivetrain losses of the vehicle at the reference velocity point v_j and part of the vehicle road load measured at the flat belt dynamometer $F_{j,\mathrm{Dyno}}$ in N;
δF_{TOM}	is the influencing quantity due to the torque measurement system TOM in N;
K_{ext}	is the correction factor for the wind tunnel method extended;
$F_{j,\mathrm{Air}}$	is the aerodynamic drag at the reference velocity point v_j in N.

Influencing quantity of the load cells δF

Each load cell of the flat belt dynamometer has a deviation of maximum $\pm 0.3\,\mathrm{N}$ in the measurement range from $0\,\mathrm{N}$ to $100\,\mathrm{N}$[12], which results in a total uncertainty of $1.2\,\mathrm{N}$ for all four load cells. It is declared as δF. The uncertainty is taken from the manufacturer specification and is therefore of type B.

Influencing quantity due to the reproducibility of the road load determination $\delta F_{j,\mathrm{Dyno}}$

The influencing quantity of three different road load measurements with braking phase executed using the test procedure SV ALT wB (see in Figure 2.28) is declared as $\delta F_{j,\mathrm{Dyno}}$. The corresponding uncertainty is described by a two-sided confidence interval of $90\,\%$ using the Student t-distribution. The deviation values of the average value of $F_{j,\mathrm{Dyno}}$ discribing the confidence interval are given in Table 4.10[30]. As these values are estimated based on independent repeated observations, this uncertainty is of type A.

With a higher number of reproducibility measurements this uncertainty could be lowered, significantly.

---

[30]For the calculation see in [13].

Table 4.10: Standard deviation $\sigma^*(F_{j,\text{Dyno}})$ estimated using the Student t-distribution of three corrected road load measurements determined using the flat belt dynamometer

v_j in $^{\text{km}}/_{\text{h}}$	20	30	40	50	60	70	80	90	100	110	120	130
$\sigma^*(F_{j,\text{Dyno}})$ in N	5.4	6.0	7.0	6.8	7.3	6.5	7.2	7.1	8.6	8.6	10.3	10.4

Influencing quantity due to the tyre inflation pressure δp_{Tyre}

The tyre inflation pressure can be adjusted with an accuracy of $\pm 0.05\,\text{bar}$[31]. As shown in subsection 4.1.4, the influence of the tyre inflation pressure can be estimated with Equation 2.24, whereby the value for the exponent is determined as -0.42 for the test vehicle F46 216i. For $p_{\text{Tyre,ISO}}$ the standard inflation pressure according to the manufacturer specification of 2.2 bar is inserted. The vehicle weight is kept constant. Afterwards, the difference in rolling resistance for the tyre inflation pressures 2.15 bar and 2.25 bar compared to the standard tyre inflation pressure of 2.2 bar are calculated. The results are stated in Table 4.11.

Table 4.11: Impact of the accuracy of the tyre pressure adjustment on the resulting rolling resistance of the test vehicle F46 216i

v_j in $^{\text{km}}/_{\text{h}}$	20	30	40	50	60	70	80	90	100	110	120	130
$\lvert F_{\text{Roll,2.2 - 2.5 bar}}\rvert$ in N	1.2	1.3	1.3	1.4	1.4	1.4	1.5	1.5	1.5	1.6	1.6	1.7
$\lvert F_{\text{Roll,2.15 - 2.2 bar}}\rvert$ in N	1.2	1.3	1.4	1.4	1.5	1.5	1.5	1.6	1.6	1.6	1.7	1.7

According to these results, it can be stated that the measured rolling resistance is determined with an uncertainty of $\pm 1.5\,\text{N}$ on average, due to the accuracy of the pressure adjustment device. This quantity is described by a rectangular PDF but is estimated due to multiple measurements and is therefore of type A.

Influencing quantity due to the force fluctuation during the measurement δF_{fluc}

The influencing quantity δF_{fluc} is referenced to fluctuations of the force measurement at each reference velocity point. The measurement time is 10 seconds with a measurement frequency of 32 Hz. This uncertainty is expressed as a standard deviation of the force and is shown in Table 4.12. It is of type A.

If there are velocity fluctuations of the flat belts, the resulting force fluctuation is included in this uncertainty. However, the velocity fluctuations of the belts are in total $\pm 0.01\,^{\text{km}}/_{\text{h}}$ for all reference velocity points v_j, as can be seen in Table E.3. In this table, the standard deviations of the belt velocity over the reference velocity points v_j are listed. Therefore, uncertainties due to velocity fluctuations of the belts can be neglected.

---

[31]Testo Industrial Services GmbH. Kalibrierschein: Reifendruckprüfer - Ewo Euroair Digital, 2018.

Table 4.12: Standard deviation $\sigma\left(F_{\text{fluc}}\right)$ of the measured force at the flat belt dynamometer

v_j in $^{\text{km}}/_{\text{h}}$	20	30	40	50	60	70	80	90	100	110	120	130
$\sigma\left(F_{\text{fluc}}\right)$ in N	0.4	0.3	0.2	0.2	0.3	0.2	0.3	0.3	0.4	0.5	1.7	0.4

Influencing quantity due to the accuracy of TOM δF_{TOM}

For each torque meter the maximum characteristic curve deviation is $0.084\,\%$ in the measurement range from $0\,\text{Nm}$ to $80\,\text{Nm}$ maximum[17]. If the dynamic tyre radius of the F46 216i of $320\,\text{mm}$ (compare Table 3.1) is assumed, the uncertainty amounts to $\pm0.84\,\text{N}$ in total for all four tyres. This is a type B evaluation.

Influencing quantities of the vehicle weighing system δm

The vehicle weight has to be determined before and after the road load measurement. Therefore, the expanded uncertainty is considered twice and has a value of $\pm7.5\,\text{kg}$ with $k_p = 2$[32]. As this information is from a calibration certificate, the uncertainty is of type B.

Influencing quantities of the measured ambient temperature δT_{Dyno} and $\delta T_{\text{Dyno,fluc}}$

The expanded uncertainty of the temperature measurement at the flat belt dynamometer amounts to $\pm0.2\,\text{K}$ with $k_p = 2$[33]. This value is taken from a calibration certificate and is therefore of type B. In addition, the temperature fluctuates with a value of $\pm0.1\,\text{K}$. However, the value of this uncertainty is in the range of the accuracy of the temperature measurement device and can be excluded from the computation, accordingly.

Finally, the expanded uncertainty U_y with $k_p = 2$ for each reference velocity point v_j in percent and referenced to the road load determined using the wind tunnel method extended is listed in Table 4.13.

Table 4.13: Expanded uncertainty U_y with $k_p = 2$ for the wind tunnel method extended for each reference velocity point v_j in percent and referenced to the road load determined using the wind tunnel method extended

v_j in $^{\text{km}}/_{\text{h}}$	20	30	40	50	60	70	80	90	100	110	120	130
U_y in %	4.3	4.4	4.5	3.8	3.6	2.8	2.7	2.4	2.5	2.2	2.5	2.1

It can be seen that the expanded uncertainty increases mostly with decreasing velocity. The maximum uncertainty, with a value of $\pm4.5\,\%$ occurs at $40\,^{\text{km}}/_{\text{h}}$. The minimum uncertainty, with a value of $\pm2.1\,\%$, can be found at the velocity of $130\,^{\text{km}}/_{\text{h}}$.

---

[32] AS-WÄGETECHNIK GMBH. Kalibrierschein: Radlastwaage Dini Argeo WWSB1500, 2018.
[33] TESTO INDUSTRIAL SERVICES GMBH. Kalibrier-Zertifikat: Wetterstation Flachband - PTU 300, 2019.

For a further analysis, the modified components of the combined standard uncertainty $u^*_{x_m,F_{\mathrm{WTMext}}}$ are discussed in the following. In Figure 4.61, the modified components of the combined standard uncertainty $u^*_{x_m,F_{\mathrm{WTMext}}}$ for the component of the aerodynamic drag determination are illustrated over the velocity points v_j. The single values are provided in Table E.4. It can be seen that the accuracy of the balance $\delta F_{x,\mathrm{acc}}$ (white bars) has the greatest impact, whereby the impact is strongly dependent on the velocity. At 20 km/h the influence is nearly zero. At the velocity of 130 km/h the impact increases to a value of about 1.5 N[34]. Furthermore, it is pointed out that the influencing factors concerning the aerodynamic drag determination are negligible compared to the influencing factors of the flat belt dynamometer (see Figure 4.62).

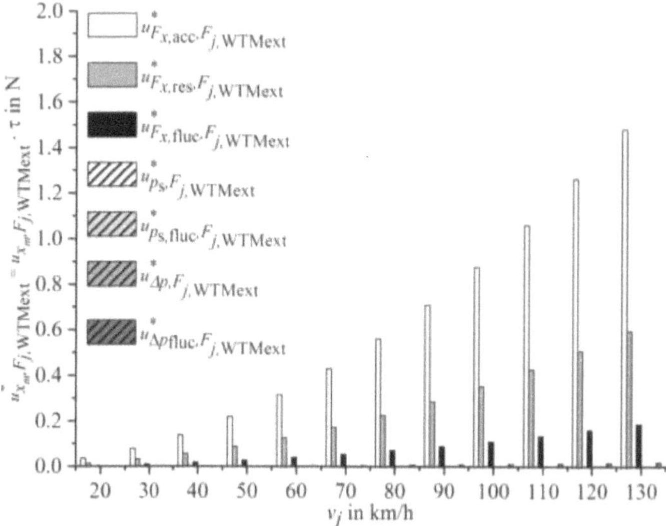

Figure 4.61: The modified components of the combined standard uncertainties $u^*_{x_m,F_{\mathrm{WTMext}}}$ for the aerodynamic drag determination using the wind tunnel method extended.

In Figure 4.62, the modified components of the combined standard uncertainties $u^*_{x_m,F_{\mathrm{WTMext}}}$ for the road load part measured at the flat belt dynamometer are plotted over the reference velocity points v_j. The values are shown in Table E.2.

It can be concluded that the uncertainty due to the reproducibility of the road load measurements, which are executed on different days $u^*_{F_{j,\mathrm{Dyno}},F_{j,\mathrm{WTMext}}}$ (dark grey bars) have the greatest impact and are mostly increasing with increasing velocity. The maximum value of 24.7 N[34] occurs at 130 km/h and the minimum value of 12.7 N[34] at 20 km/h. These

---

[34]This value is not the real value for the component of the combined standard uncertainty. Instead, it is multiplied with the factor τ (see Equation 4.14).

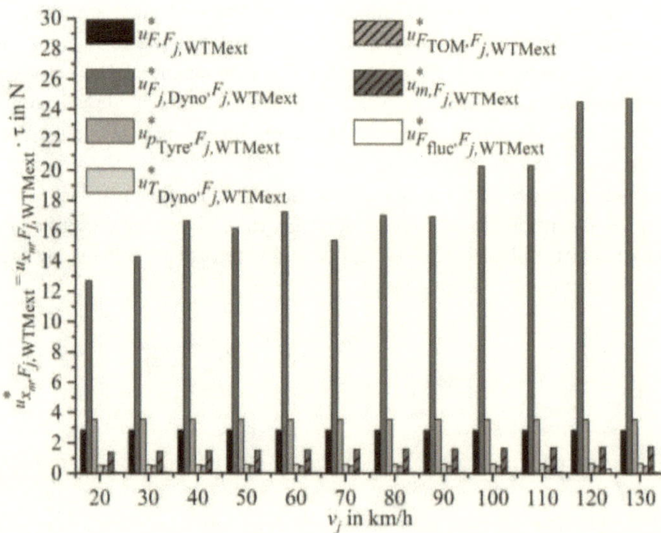

Figure 4.62: The modified components of the combined standard uncertainty $u^*_{x_m,F_\text{WTMext}}$ for the flat belt dynamometer using the wind tunnel method extended.

high values are mainly caused by the low number of reproducibility measurements (three measurements).

4.4.3 AEROLAB method

Generally, the road load according to the AEROLAB method can be determined using Equation 3.8. However, the part $F_{j,\text{WT,woB}}$ is already corrected to reference conditions using the correction equations, similar to the ones used for the coastdown method (see subsection 3.1.3). Only the wind correction w_1 is set to zero, as there is no additional tailwind and headwind in the wind tunnel. Considering that the correction terms for the coastdown method need a polynomial regression, the error calculation has to be performed again after this polynomial regression.

The relation between the equation for the road load measured in the AEROLAB wind tunnel $F_{j,\text{WT,woB}}$ and the influencing quantities is given as:

$$F_{j,\text{WT,woB}} = \left((f_0 + \delta f_0 + \delta F_{\text{fluc}} - K_1) + (f_1 + \delta f_1) \cdot v_j\right) \cdot$$
$$\left(1 + K_0 \cdot \left(\overline{T}_{\text{Amb}} + \delta T_{\text{Amb}} + \delta T_{\text{Amb,fluc}} - T_0\right)\right) + K_2 \cdot (f_2 + \delta f_2) \cdot v_j^2 \quad (4.23)$$

with

$$K_2 = \frac{\overline{T}_{\text{Amb}} + \delta T_{\text{Amb}} + \delta T_{\text{Amb,fluc}}}{293\,\text{K}} \cdot \frac{100\,\text{kPa}}{\overline{p}_{\text{Amb}} + \delta p_{\text{Amb}} + \delta p_{\text{Amb,fluc}}} \quad (4.24)$$

where:

$F_{j,\text{WT,woB}}$	is the vehicle road load without residual brake forces measured in the wind tunnel in N (see subsection 3.1.3);
f_0	is the constant term of the road load coefficients in N;
δf_0	is the influencing quantity of the uncorrected road load coefficient f_0 in N;
δF_{fluc}	is the influencing quantity due to force fluctuations during the force measurement with the load cells integrated in the vehicle fixation system in N;
K_1	is the test mass correction factor (see Equation 2.47);
f_1	is the coefficient of the first order term of the road load coefficients in $\text{N}/(\text{m/s})$;
δf_1	is the influencing quantity of the uncorrected road load coefficient f_1 in $\text{N}/(\text{m/s})$;
v_j	is the reference velocity point in m/s;
K_0	is the correction factor for rolling resistance (and drivetrain losses) in K^{-1} (see Equation 2.46);
$\overline{T}_{\text{Amb}}$	is the arithmetic average ambient atmospheric temperature in K;
δT_{Amb}	is the influencing quantity of the measured ambient temperature T_{Amb} in K;
$\delta T_{\text{Amb,fluc}}$	is the influencing quantity due to fluctuations of the ambient temperature T_{Amb} during the measurement in K;
T_0	is the reference atmospheric temperature of $20\,^\circ\text{C}$ in K;
K_2	is the air resistance correction factor;
$\overline{p}_{\text{Amb}}$	is the arithmetic average ambient atmospheric pressure in kPa;
δp_{Amb}	is the influencing quantity of the measured ambient pressure p_{Amb} in kPa;
$\delta p_{\text{Amb,fluc}}$	is the influencing quantity due to fluctuations of the ambient pressure p_{Amb} during the measurement in kPa;
f_2	is the coefficient of the second order term of the road load coefficients in $\text{N}/(\text{m/s})^2$;
δf_2	is the influencing quantity of the uncorrected road load coefficient f_2 in $\text{N}/(\text{m/s})^2$.

The equations for K_0 and K_1 are equivalent to the Equations 4.10 and 4.11. They are already defined for the error calculation of the coastdown method (see subsection 4.4.1). In addition to that, the road load according to the AEROLAB method is described by the following equation with the additional influencing quantities:

$$F_{j,\text{AM}} = F_{j,\text{Brake}} + \delta F_{\text{Brake}} + F_{j,\text{WT,woB}} + \delta F_{j,\text{WT,woB}} \qquad (4.25)$$

where:

$F_{j,\text{AM}}$ is the road load of a vehicle at the reference velocity point v_j determined using the AEROLAB method in N (see Equation 3.8);

$F_{j,\text{Brake}}$ is the residual brake force at the reference velocity point v_j determined at the flat belt dynamometer in N (see Equation 3.7);

δF_{Brake} is the influencing quantity due to the determination procedure of the residual brake forces in N;

$\delta F_{j,\text{WT,woB}}$ is the influencing quantity of the repeatability of the corrected road load determined measured in the AEROLAB wind tunnel at different days in N.

The influencing quantities are explained in the following.

Influencing quantities of the uncorrected road load coefficients δf_0, δf_1 and δf_2

According to the AEROLAB method, the road load is measured with two load cells, which are integrated into the vehicle fixation system (see Figure 3.8). Furthermore, the road load is determined using the test procedure SV ALT woB (see Figure 3.9). Therefore, related to the road load coefficients, the uncertainty is mainly defined by the accuracy of the load cells. The load cells have an accuracy of $0.6\,\text{N}$[13]. As a consequence, the uncertainty of the road load coefficient f_0 amounts to $\pm 1.2\,\text{N}$ (two times $0.6\,\text{N}$). The uncertainties for the other two road load coefficients f_1 and f_2 are zero, as it is assumed that the uncertainty due to the load cells only results in an offset shift of the measured forces. Due to the rectangular PDF of the accuracy, this uncertainty is a type B evaluation.

Influencing quantity of the force fluctuations measured with the load cells δF_{fluc}

Similar to the flat belt dynamometer, the losses of the vehicle in the AEROLAB wind tunnel are measured with two load cells, which are integrated into the vehicle restraint system. Fluctuations in the force measurement can occur because of uncertainties of the wind velocity and belt velocity, as well as possible pulsating flow separations at the vehicle and possible deformations of the external shell of the vehicle. This uncertainty is accounted for by the standard deviation of the measured force at each reference velocity, as listed in Table 4.14. The measurement time is 10 seconds with a measurement frequency of $128\,\text{Hz}$. It is a type A uncertainty, as the standard deviations are used.

Table 4.14: Standard deviation $\sigma\left(F_{\text{fluc}}\right)$ of the measured force during the force measurement in the AEROLAB wind tunnel

v_j in $^{\text{km}}/_{\text{h}}$	20	30	40	50	60	70	80	90	100	110	120	130
$\sigma\left(F_{\text{fluc}}\right)$ in N	2.8	1.7	1.3	2.1	2.0	2.5	3.3	4.4	4.2	7.2	5.0	7.3

Influencing quantities of the corrected road load $\delta F_{j,\text{WT,woB}}$

However, the uncertainty previously described considers only one measurement. Therefore, $\delta F_{j,\text{WT,woB}}$ is defined as the standard deviation of six single road load measurements in the AEROLAB wind tunnel. The standard deviation is given in Table 4.15. As the uncertainty is determined using a standard deviation, it is of type A.

Table 4.15: Standard deviation $\sigma\left(F_{j,\text{WT,woB}}\right)$ of the corrected road load curves of six single road load measurements in the AEROLAB wind tunnel

v_j in $^{\text{km}}/_{\text{h}}$	20	30	40	50	60	70	80	90	100	110	120	130
$\sigma\left(F_{j,\text{WT,woB}}\right)$ in N	3.4	4.2	5.0	5.5	5.7	5.7	5.5	5.0	4.4	3.9	3.9	4.9

Influencing quantity of the residual brake forces $\delta F_{j,\text{Brake}}$

The residual brake forces are determined using the flat belt dynamometer. The corresponding uncertainty is defined as the standard deviation, which is given in Figure 4.28 (light grey open circles). This uncertainty is of type A.

Influencing quantities of the vehicle weighing system δm

Similar to the other methods, the vehicle weight has to be determined before and after of the road load measurements. Therefore, the uncertainty has to be considered twice and has a value of $\pm 7.5\,\text{kg}$ with $k_p = 2^{32}$. It is a type B evaluation.

Influencing quantities of the measured ambient conditions

The ambient temperature in the AEROLAB wind tunnel is measured with a platinum resistance thermometer (Pt100). According to DIN EN 60751, the accuracy at $20\,^{\circ}\text{C}$ is $\pm 0.19\,\text{K}$ $\left(\delta T_{\text{Amb}}\right)$[35]. The ambient pressure is determined with an accuracy of $\pm 0.012\,\text{Pa}$ $\left(\delta p_{\text{Amb}}\right)$[36]. Both quantities are of type B. During the road load determination in the wind tunnel, the temperature fluctuates with a standard deviation of $\pm 0.19\,\text{K}$ $\left(\delta T_{\text{Amb,fluc}}\right)$ and the ambient pressure with a standard deviation of $\pm 0.3845\,\text{Pa}$ $\left(\delta p_{\text{Amb,fluc}}\right)$. These uncertainties are both of type A.

---

[35]DIN EN 60751:2009-05. Industrielle Platin-Widerstandstehermometer und Platin-Thermosensoren (IEC 60751:2008); Deutsche Fassung EN 60751:2008, 2008.
[36]MKS INSTRUMENTS INC.. MKS Type 690A Absolute and Type 698A: Differential High Accuracy Pressure Transducers, 2009.

Finally, the expanded uncertainty U_y with $k_p = 2$ for each reference velocity point v_j in percent and referenced to the road load determined using the AEROLAB method are listed in Table 4.16.

Table 4.16: Expanded uncertainty U_y with $k_p = 2$ for the AEROLAB method after the polynomial regression for each reference velocity point v_j in percent and referenced to the road load determined using the AEROLAB method

v_j in $^{km}/_h$	20	30	40	50	60	70	80	90	100	110	120	130
U_y in %	2.6	2.5	2.4	2.3	2.0	1.7	1.5	1.2	1.0	0.9	0.8	0.8

It can be seen that the uncertainty is mostly increasing with decreasing velocity similar to the wind tunnel method extended. The maximum uncertainty, with a value of 2.6 %, occurs at $20\,^{km}/_h$. The minimum uncertainty of 0.8 % occurs at $120\,^{km}/_h$ and $130\,^{km}/_h$. For further analysis, the modified components of the combined standard uncertainty $u^*_{x_m,F_{AM}}$ are plotted over the reference velocity points v_j in Figure 4.63. In addition to that, the values of the modified components of the standard uncertainty are listed in Table E.5.

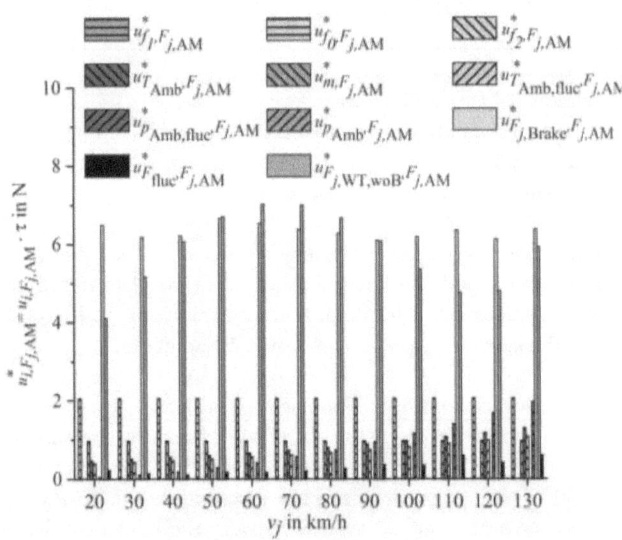

Figure 4.63: The modified components of the combined standard uncertainty $u^*_{x_m,F_{AM}}$ for the input quantities δx_m of the AEROLAB method.

From this illustration, it is apparent that the determination of the residual brake forces $u^*_{F_{j,Brake},F_j,AM}$ (light grey bars) and the influence of several measurements $u^*_{F_{j,WT,woB},F_j,AM}$ (dark grey bars) have the largest impact on the expanded uncertainty. In contrast, the other influencing quantities can almost be neglected. Furthermore, it can be seen that all uncertainties are nearly independent of the velocity, which explains the reduction of the expanded uncertainty U_j with increasing velocity v_j.

Note: All values for the components of the combined uncertainty $u_{x_m,F_j,\mathrm{AM}}$ are multiplied with the factor τ as explained in Equation 4.14. The result is given as the modified component of the combined uncertainty $u^*_{x_m,F_j,\mathrm{AM}}$.

4.4.4 Summary

The major findings on the error calculation depending on the chosen road load determination method are summarized in the following.

<u>Coastdown method:</u>

- The largest uncertainty for the error calculation 'After polynomial regression' occurs for the uncertainties of the uncorrected road load coefficients f_1 and f_2. Due to their velocity dependency, these uncertainties also include the uncertainty of the velocity determination via the GPS device.

- However, the road load coefficients are not independent quantities, therefore the correlation term has to be taken into account (see Equation 2.90). Due to the mainly negative correlation coefficients the expanded uncertainty is reduced.

- The influences of the modified components of the combined standard uncertainty $u^*_{T_{\mathrm{Amb}},F_j,\mathrm{CDM}}$, $u^*_{p_{\mathrm{Amb}},F_j,\mathrm{CDM}}$, $u^*_{v_{\mathrm{Wind}},F_j,\mathrm{CDM}}$ and $u^*_{m,F_j,\mathrm{CDM}}$ are negligible.

<u>Wind tunnel method extended:</u>

- Considering the aerodynamic drag determination, the accuracy of the balance has the greatest influence on the expanded uncertainty. The maximum value of $u^*_{F_{x,\mathrm{acc}},F_j,\mathrm{WTMext}}$ is about $1.5\,\mathrm{N}$[37] at the reference velocity point of $130\,\mathrm{km/h}$. The other modified components of the combined standard uncertainty $u^*_{x_m,F_j,\mathrm{WTMext}}$ have a value less than $1\,\mathrm{N}$[37]. Compared to the influencing quantities of the flat belt dynamometer part the influencing quantities concerning the aerodynamic drag determination are negligible.

- Considering the road load measurement part at the flat belt dynamometer, the uncertainty of the reproducibility of the road load determination at different days $u^*_{F_j,\mathrm{Dyno},F_j,\mathrm{WTMext}}$ has the largest impact with a maximum value of $24.7\,\mathrm{N}$[37] at the reference velocity point of $130\,\mathrm{km/h}$. This is mainly caused by the low number of three reproducibility measurements. If the number of reproducibility measurements is increased, this uncertainty could be lowered, significantly. For comparison, the reproducibility for the AEROLAB method is made with six measurements.

---

[37]This value is not the real value for the component of the combined standard uncertainty. Instead, it is multiplied with the factor τ (see Equation 4.14)

- Due to the investigation of road load measurements with and without residual brake forces (see Figure 4.28), it can be assumed that the uncertainty of the road load determined at the flat belt dynamometer is also mainly related to the residual brake forces.

<u>AEROLAB method:</u>

- The expanded uncertainty of the AEROLAB method is mainly characterizied by the determination of the residual brake forces and the reproducibility of the road load measurements without the residual brake forces in the wind tunnel.

- As in the case of the wind tunnel method extended, the uncertainties due to the influencing factors related to the ambient conditions are negligible.

Finally, the expanded uncertainty U_y with $k_p = 2$ for the coastdown method (after polynomial regression) (black bars), the wind tunnel method extended (grey bars) and the AEROLAB method (light grey bars) are plotted over the reference velocity points v_j as a bar chart.

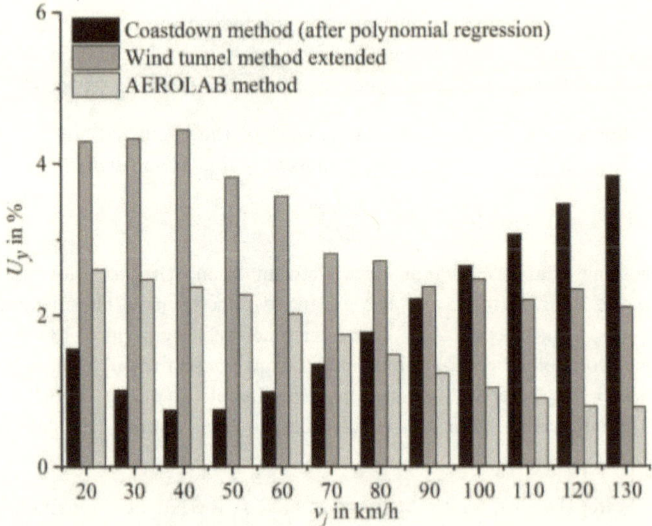

Figure 4.64: Expanded uncertainty U_y with $k_p = 2$ for the coastdown method (after polynomial regression) (black bars), the wind tunnel method extended (grey bars) and the AEROLAB method (light grey bars).

It can be concluded that the uncertainty is decreasing with the velocity for both the wind tunnel method extended and the AEROLAB method and is increasing for the coastdown method. Moreover, the maximum value for the wind tunnel method extended is 4.4 %

at the velocity of 40 km/h. At velocities greater than 90 km/h the expanded uncertainty for this method is lower than for the coastdown method. A reason for the high values of the expanded uncertainty of the wind tunnel method extended is the uncertainty due to the low number of reproducibility measurements. The expanded uncertainty for the AEROLAB method decreases from 2.6 % at 20 km/h to 0.8 % at 130 km/h. In contrast, the lowest uncertainty of the coastdown method occurs at 40 km/h and 50 km/h and has value of about 0.7 %. The maximum value of about 3.8 % occurs at a velocity of 130 km/h.

CHAPTER 5

Conclusion and need for further research

The determination of the road load is the first and a very important step for the total type approval procedure, as stated in the introduction of this **book** (see chapter 1). Therefore, an extensive literature review on the composition of the vehicle road load and the different methods to determine the road load, which are allowed since the inception of WLTP, was performed and summarised at first. Only with the knowledge of the several influencing factors on the road load and of the different determination methods, it was possible to develop a new method to determine the road load of a ve-hicle also in the wind tunnel. The development of this new method is the focus of the **book**.

Therefore, the so-called wind tunnel method described in the GTR No. 15 was investigated in detail, initially. According to this method, the aerodynamic drag is measured in a wind tunnel and the remaining components (rolling resistance and drivetrain losses) are determined using a flat belt d ynamometer. To be able to assign the investigated sensitivities and influencing factors to one single component, a measurement method using a custom-built torque meter was developed by Untermaierhofer, Petz and Vogeler (see [72]). With this measurement method, called TOM, it is possible to separate the flat b elt dynamometer measurement result into its two components.

The first result of the flat belt dynamometer investigations is that the cooling fan, which stands in front of the vehicle, has no influence on the force m easurement. T his is an important result as the aerodynamic drag component of the road load has to be determined only in the wind tunnel.

Furthermore, the different test procedures defined in the GTR No. 15 [10] were investigated. There was no difference in the measured road load depending on the chosen measurement phase (stabilized or descending velocity), however depending on the chosen warm-up phase (standard or alternative warm-up) a difference could be demonstrated. If the vehicle is driven by its own engine (standard warm-up), the resulting losses are lower due to higher transmission oil temperatures. But neither the type of the warm-up phase nor the chosen measurement phase have an influence on the tyre tread temperatures.

In addition, in the literature [26, 30] an estimation is given, which describes the influence of the tyre inflation pressure on the resulting rolling resistance. For this estimation, the rolling resistance is measured using a test drum according to ISO 8767/ISO 18164 [75, 76]. Therefore, the transfer of this assumption to the rolling resistance measured on the flat belt dynamometer was investigated. It was found that the estimation is similar for both measurement procedures at the flat belt dynamometer and on the test drum. However, it was shown that the influence of the tyre inflation pressure is not constant for all investigated tyre inflation pressures and velocities. Therefore, the standard deviation of the computed exponent value, which is used for the estimation, has the same magnitude as the value itself. Considering this, it is only an approximate estimation.

Moreover, it was shown that the tolerance of the vehicle position on the flat belts of the test bench defined in the GTR No. 15 [10] results in a difference of up to 20 N referenced to the rolling resistance. The deviation can be explained by an additional toe at the rear axle, if the rear axle is not centered at the flat belts.

In a next step, the influence of the ambient temperature change from 23 °C to 15 °C and to 10 °C on the rolling resistance and on the drivetrain losses was determined. In the case that the ambient temperature at the test bench is decreased from 23 °C to 10 °C, the tyre tread temperatures and the transmission oil temperature decrease between 10 °C and 11 °C. This temperature decrease leads to an increase of the rolling resistance of about 13 %. Accordingly, these results cannot confirm the statement made in [19] that a temperature decrease of the tyre treads of about 10 % leads to a rolling resistance increase of about 43 % in the investigated temperature range from 41.5 °C to 46 °C. Besides the rolling resistance, also the drivetrain losses increase due to the decreased ambient temperature and the decreased transmission oil temperature. The drivetrain losses at an ambient temperature of 10 °C increased by about 14 % on average as compared to the losses at 23 °C.

After the detailed analysis of the measurement method at the flat belt dynamometer, the road load determined using the flat belt dynamometer in combination with the BMW Group wind tunnel was compared with the road load measured on a proving ground using the coastdown method. However, for this comparison the wind tunnel method according to GTR No. 15 was extended with an additional correction factor K_{ext}. This correction factor considers that the rolling resistance for different wheel loads on asphalt is on average 20 % to 30 % larger as compared to a surface coated with Safety Walk^{TM} [38], which is used as surface coating of the belts at the flat belt dynamometer. This newly developed method is called wind tunnel method extended. The additional correction factor K_{ext} with a value of 1.34 was determined in such a way that the averaged difference in cycle energy demand between the wind tunnel method extended and the coastdown method calculated for five different vehicles was minimal. Afterwards, the newly developed wind tunnel method extended was verified with further eight vehicles. It was shown that the maximum difference in cycle energy demand related to the WLTC is 3.5 % for the F21 120iA. Therefore, for all investigated vehicles the difference in cycle energy demand lies in the range of ±5%, which is also an acceptance criterion for the wind tunnel method according to GTR No. 15 [10].

As already stated at the beginning, the focus of this thesis is the development of a method to determine the total road load of a vehicle in the wind tunnel. It was shown that with the newly developed AEROLAB method it is possible to determine the road load excluding the residual brake forces in the AEROLAB wind tunnel of the BMW Group. The residual brake forces were determined using the flat belt dynamometer. The AEROLAB method was verified in total with three different vehicles. The maximum difference in cycle energy demand is 3.2 % for the MINI F60 Countryman. The above stated acceptance criterion for the wind tunnel method according to GTR No. 15 is also fulfilled for the AEROLAB method with the investigated vehicles.

Although in the AEROLAB wind tunnel a steel belt is used, there is no further correction of the rolling resistance applied, as for the wind tunnel method extended, to reach the road load level on a proving ground. And according to [39, 40], the rolling resistance on a steel surface is even lower than on a surface coated with Safety Walk^{TM}. Therefore, several differences between the wind tunnel method extended and the AEROLAB method were examined and rated. It was shown that the additional wheel ventilation resistance and the increased rolling resistances and drivetrain losses, due to lower tyre tread [60] and transmission oil temperatures [60], are mainly responsible for the higher road load measured using the AEROLAB method. In contrast to the AEROLAB method, the wheel ventilation resistance is an internal force, if the aerodynamic drag is measured using the test procedure defined in the GTR No. 15 wind tunnel method and in the wind tunnel method extended, and is therefore no component of the road load determined using the wind tunnel method extended. Furthermore, in the AEROLAB wind tunnel a better cooling effect of the tyres and of the drivetrain is assumed due to the $14\,\mathrm{m}^2$ nozzle in front of the vehicle and due to the rolling-road-single-belt system. On the other hand, at the flat belt dynamometer only a cooling fan with an area of about $0.3\,\mathrm{m}^2$ was used and each vehicle tyre stands on a single flat belt. In addition, there is no further centre belt, to simulate an underbody airstream and, therefore, to cool the tyres and the drivetrain more effectively [60]. Moreover, it could be shown that the different force measurement system used for the AEROLAB method and the steel surface of the rolling-road-single-belt system would result in a lower road load as compared to the road load measured using the flat belt dynamometer. However, the influence of these factors on the road load are much lower than the influence of the additional wheel ventilation resistance and the one of the lower tyre tread temperatures and transmission oil temperature.

Finally, for alle three investigated methods - coastdown method, wind tunnel method extended and AEROLAB method - an error calculation according to GUM was performed. The greatest uncertainty in the velocity range greater than $90\,\mathrm{km/h}$ could be found in the coastdown method. The greatest expanded uncertainty of the coastdown method with $k_p = 2$ with a value of about 3.8 % occurs at a velocity of $130\,\mathrm{km/h}$. Furthermore, the expanded uncertainty of this method mainly increases with increasing velocity. In contrast, for the wind tunnel method extended and the AEROLAB method, the expanded uncertainty with $k_p = 2$ increases with decreasing velocity. The greatest expanded uncertainty was about 4.4 % for the wind tunnel method extended. This is mainly caused by the low number of reproducibility measurements. In the case of a higher number of measurements this uncertainty could be lowered. In contrast, the maximum expanded uncertainty for the AEROLAB method was about 2.6 % at a velocity of 20 km/h. The minimum value of about 0.8 % occurs at the velocities of 120 km/h and 130 km/h. The largest uncertainties for the AEROLAB method could be attributed to the reproducibility of the road load determination at different days and the determination of the residual brake forces.

It can be concluded that the determination of the vehicle road load in the AEROLAB wind tunnel of the BMW Group is possible. However, in this study the residual brake forces were externally determined at the flat belt dynamometer. For the determination of the total road load in the AEROLAB wind tunnel, an exhaust system has to be integrated into the wind tunnel, so that the measurement can be executed with a running engine. In addition, the wind tunnel belt system must be able to perform a driving simulation, so that the vehicle can accelerate to 80 km/h and can then decelerate to 20 km/h as defined for the braking phase. Finally, a further verification with vehicles containing automatic transmissions is necessary, as up to now only vehicles with a manual transmission have been used to develop the AEROLAB method. Overall, due to the large number of differences between the wind tunnel method extended and the AEROLAB method, these influencing factors should be investigated with further vehicles, as they have a significant effect on the resulting road load. Furthermore, it was shown that the vehicle fixation system results in an increased aerodynamic drag. Therefore, a compensation term, which is valid for different vehicle types, has to be developed. In addition to that, the wind tunnel method extended should also consider a further correction term for the temperature dependence on drivetrain losses. Therefore, a detailed investigation into the temperature influence on the drivetrain losses of manual and automatic transmissions is needed.

Finally, the resulting road loads of the wind tunnel method extended and of the AERO-LAB method were compared with the results of the coastdown method. However, the coastdown method is strongly influenced by environmental conditions, e.g. wind and atmospheric temperature, but also by steering interventions of the driver. Therefore, further research should include the investigation of these influencing factors and their impact on the resulting road load using the coastdown method. In particular, extensive research on environmental conditions influencing the transmission oil temperature and thus the drivetrain losses is needed. Moreover, different cooling behaviours of the tyres and also of the drivetrain on the test benches and also during road load measurements on the road should be investigated in more detail. Last but not least, a surface for test benches is needed, which will simulate the correct rolling resistance characteristics as they exist in normal road conditions.

APPENDIX A

Wind tunnel method

A.1 Modified test procedure

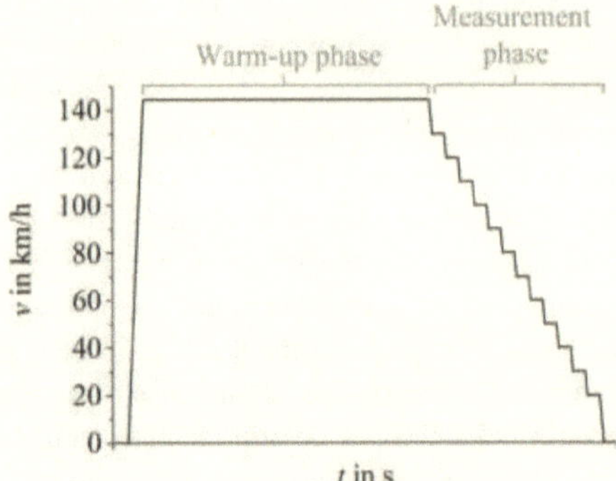

Figure A.1: Modified test procedure without braking phase SV ALT woB with a stabilization time of 4 seconds prior to each measurement point (adapted by permission from Springer Nature Customer Service Centre GmbH: Springer Nature, 18. Internationales Stuttgarter Symposium [8], 2018).

A.2 Test vehicles

In the following two tables (Table A.1 and Table A.2), further information about the vehicles, which are used to develop and to verify the wind tunnel method extended, are given.

Table A.1: Characteristics of the test vehicles (Part I)

Derivate	Model	Transmission type	Wheel drive	Tyre manufacturer	Tyre size
G11	725dA	AT	RWD	Michelin	FA: 225/60 R17 99 Y NL STD RA: 225/60 R17 99 Y NL STD
F46	216i	MT	FWD	Pirelli	FA: 225/45 R18 95 V XL STD RA: 225/45 R18 95 V XL STD
F33	430i	MT	RWD	Continental	FA: 225/45 R18 95 V XL RFT RA: 225/45 R18 95 V XL RFT
F21	120i	AT	RWD	Continental	FA: 225/45 R17 91 H NL RFT RA: 225/45 R17 91 H NL RFT
G15	M850iA	AT	AWD	Michelin	FA: 245/35 R20 95 Y XL STD RA: 275/30 R20 97 Y XL STD
G20	330iA	AT	RWD	Michelin	FA: 225/40 R19 93 Y XL STD RA: 255/35 R19 96 Y XL STD
G30	530e iPerformance	AT	RWD	Michelin	FA: 245/40 R19 98 Y XL RFT RA: 275/35 R19 100 Y XL RFT

Table A.2: Characteristics of the test vehicles (Part II)

Derivate	Model	Transmission type	Wheel drive	Tyre manufacturer	Tyre size
G07	X7 M50dA	AT	RWD	Pirelli	FA: 285/45 R21 113 Y XL RFT RA: 285/45 R21 113 Y XL RFT
J29	SPX30iA	AT	RWD	Hankook	FA: 225/50 R17 Y XL STD RA: 255/45 R17 98 Y NL STD
G12	750Ld	AT	AWD	Pirelli	FA: 245/50 R18 H NL RFT RA: 245/50 R18 H NL RFT
G14	M850iA	AT	AWD	Michelin	FA: 245/35 R20 95 Y XL RFT RA: 275/30 R20 97 Y XL RFT
F60	Countryman	MT	FWD	Pirelli	FA: 225/50 R18 NL RFT RA: 225/50 R18 NL RFT
G30	518d	MT	RWD	Continental	FA: 225/55 R17 97 H NL RFT RA: 225/55 R17 97 H NL RFT

A.3 Equilibrium of forces at the flat belt dynamometer

In Figure A.2 the equilibrium of forces at the flat belt dynamometer is shown, in the case that the vehicle is mounted horizontally at the tow hooks of the vehicle and that the vehicle cannot move in x direction.

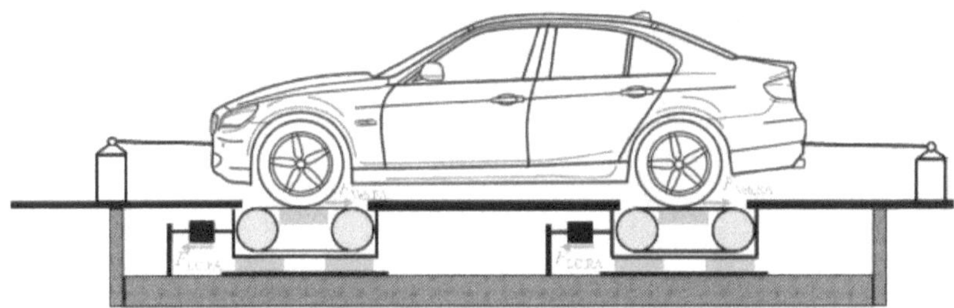

Figure A.2: Equilibrium of forces using the flat belt dynamometer (adapted by permission from Springer Nature Customer Service Centre GmbH: Springer Nature, 18. Internationales Stuttgarter Symposium [8], 2018).

Furthermore, due to the restraint and measurement system, it is ensured that there is no aerodynamic drag in the measurement results (see the results in subsection 4.1.1). Additionally, if the restraint system of the test bench is mounted at the tow hooks of the vehicle horizontally, it can be assumed that the forces measured at the load cells contain only the rolling resistance and the drivetrain losses of the vehicle:

$$F_{\mathrm{LC,FA}} = F_{\mathrm{Veh,FA}} \tag{A.1}$$

and

$$F_{\mathrm{LC,RA}} = F_{\mathrm{Veh,RA}} \tag{A.2}$$

where:

$F_{\mathrm{LC,FA}}$	is the sum of the measured forces at the both load cells (LC) of the front WDUs in N;
$F_{\mathrm{LC,RA}}$	is the sum of the measured forces at the both load cells (LC) of the rear WDUs in N;
$F_{\mathrm{Veh,FA}}$	are the vehicle losses at the front axle of a vehicle in N;
$F_{\mathrm{Veh,RA}}$	are the vehicle losses at the rear axle of a vehicle in N.

A.4 Dynamometer load setting

For the simulation of the aerodynamic drag force of the vehicle the dynamometer load setting F_d is defined as [10]:

$$F_{j,\mathrm{d}} = a_\mathrm{d} + b_\mathrm{d} \cdot v_j + c_\mathrm{d} \cdot v_j^2 \tag{A.3}$$

with:

$$a_\mathrm{d} = 0 \tag{A.4}$$

$$b_\mathrm{d} = 0 \tag{A.5}$$

and

$$c_\mathrm{d} = (c_\mathrm{w} \cdot A_x) \cdot \frac{\rho_0}{2} \tag{A.6}$$

where:

$F_{j,\mathrm{d}}$	is the dynamometer load setting in N;
a_d	is the first dynamometer set coefficient in N;
b_d	is the second dynamometer set coefficient in $\mathrm{N}/\mathrm{(m/s)}$;
c_d	is the third dynamometer set coefficient in $\mathrm{N}/\mathrm{(m/s)}^2$;
v_j	is the reference velocity point in $\mathrm{m/s}$;
c_w	is the aerodynamic drag coefficient;
A_x	is the frontal area of the vehicle in m^2;
ρ_0	is the dry air density and is defined as 1.189 $\mathrm{kg/m^3}$.

The equivalent inertia of the dynamometer is the test mass of the vehicle [10].

A.5 Difference in the road load determination between the theoretical integration and the simplification

In section 2.1, the theoretical correct calculation of the road load with the aid of an integration is introduced. To simplify the calculation, in GTR No. 15 [10] the calculation for the coastdown method has been simplified (see Equation 2.36). Therefore, in Figure A.3 the resulting road loads determined using both methods are illustrated over the reference velocity points v_j. Additionally, the difference between these two curves is plotted using the right axis. For the calculation the assumed coastdown times Δt_j are given in Table A.3. The assumed mass of the vehicle is 2000 kg. The results for the road load coefficients for both methods are stated in Table A.4.

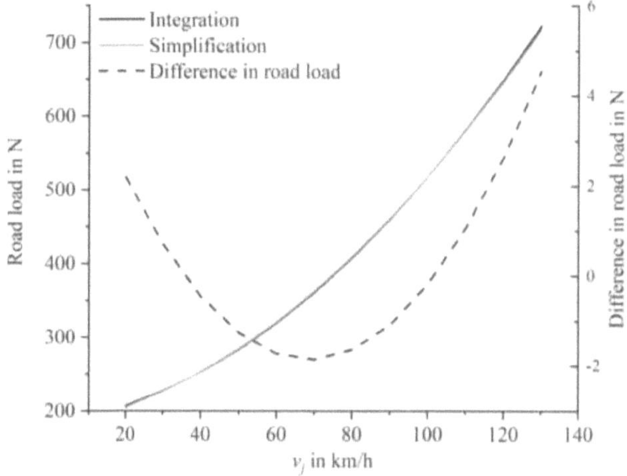

Figure A.3: Difference between the road load determined by an integration (see subsection 2.1) and by the simplification defined in the GTR NO. 15 [10].

It is pointed out that the maximum difference is 4.5 N at the velocity of 130 km/h. To investigate the effect of this difference in the road load, the difference in cycle energy demand between these two different calculation methods is determined based on Equation 2.71. The results are given in Table A.5.

The maximum difference in cycle energy demand is -0.1 % for the ARTEMIS driving cycle. Therefore, it is shown that there is no significant difference between these two methods and that both methods can be used equivalently.

Table A.3: Assumed coastdown times Δt_j

v_j in $^{km}/_h$	Δt_j in s
20	26.7
30	24.4
40	22.3
50	19.8
60	17.1
70	15.7
80	13.6
90	12.0
100	10.6
110	9.5
120	8.6
130	7.8

Table A.4: Road load coefficients f_0, f_1 and f_2 for both calculation methods (integration and simplification)

Method	f_0 in N	f_1 in $^N/_{(m/s)}$	f_2 in $^N/_{(m/s)^2}$
Integration	183.1	2.3564	0.34850
Simplification	176.9	3.2005	0.32640

Table A.5: Differences in cycle energy demand ϵ between the cycle energy demands using the road load calculated with an integration (see section 2.1) and with the simplification defined in GTR No. 15 [10]

	WLTC	ARTEMIS
ϵ in %	0.0	-0.1

APPENDIX B

Flat belt dynamometer

B.1 Separating rolling resistance and drivetrain losses

The following figure shows the ratio between the drivetrain losses $F_{j,\text{Drive}}$ and the rolling resistance $F_{j,\text{Roll}}$ over the reference velocity points v_j for the test vehicle F46 216i measured using the flat belt dynamometer. For the determination the test procedure SV ALT woB (see Figure A.1) is used.

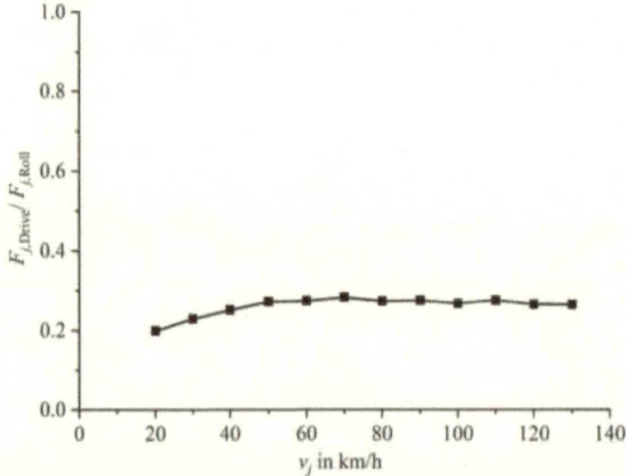

Figure B.1: Ratio between the drivetrain losses $F_{j,\text{Drive}}$ and the rolling resistance $F_{j,\text{Roll}}$ over the reference velocity points v_j for the test vehicle F46 216i measured using the flat belt dynamometer.

B.2 Tyre inflation pressure

In the following tables the computed exponents $e_{j,p}$ for the front and for the rear axle of the test vehicles G11 725dA and F46 216i are given.

Table B.1: Exponents e_{v_j} at each reference velocity point v_j for the front and rear axle with $p_{\mathrm{Tyre,Std}} = 2.2\,\mathrm{bar}$ of the test vehicle G11 725dA and the corresponding coefficients of the determination r^2

v_j in $\mathrm{km/h}$	Front axle		Rear axle	
	e_{v_j}	r^2	e_{v_j}	r^2
20	-0.448	0.9920	-0.314	0.9186
30	-0.446	0.9909	-0.324	0.9373
40	-0.441	0.9904	-0.322	0.9231
50	-.0446	0.9930	-0.313	0.9172
60	-0.441	0.9919	-0.317	0.9023
70	-0.445	0.9924	-0.312	0.8867
80	-0.456	0.9918	-0.309	0.8703
90	-0.469	0.9918	-0.311	0.8621
100	-0.476	0.9915	-0.283	0.7817
110	-0.491	0.9928	-0.423	0.8974
120	-0.514	0.9915	-0.355	0.7247
130	-0.526	0.9905	-0.456	0.8739

Table B.2: Exponents e_{v_j} at each reference velocity point v_j for the front and rear axle with $p_{\mathrm{Tyre,Std}} = 2.2\,\mathrm{bar}$ of the test vehicle F46 216i and the corresponding coefficients of the determination r^2

v_j in $\mathrm{km/h}$	Front axle		Rear axle	
	e_{v_j}	r^2	e_{v_j}	r^2
20	-0.386	0.9426	-0.355	0.9943
30	-0.402	0.9695	-0.371	0.9935
40	-0.412	0.9807	-0.372	0.9935
50	-0.409	0.9882	-0.380	0.9932
60	-0.416	0.9936	-0.387	0.9929
70	-0.422	0.9985	-0.398	0.9925
80	-0.437	0.9999	-0.405	0.9923
90	-0.445	0.9995	-0.411	0.9922
100	-0.455	0.9981	-0.420	0.9924
110	-0.467	0.9962	-0.423	0.9912
120	-0.480	0.9936	-0.433	0.9903
130	-0.502	0.9924	-0.441	0.9902

APPENDIX C

Development and verification of the wind tunnel method extended

---

In Figure C.1 the differences in road load determined with the wind tunnel method extended $F^c_{j,\mathrm{WTMext}}$ related to the coastdown method $F^c_{j,\mathrm{CDM}}$ are plotted over the reference velocity points v_j for the eight verification vehicles (see subsection 4.2.2).

In this figure, the vehicles, which are illustrated with a black curve line, have a greater difference in cycle energy demand calculated on the basis of the WLTC than calculated on the basis of the ARTEMIS European driving cycle. It can be seen that all these vehicles (G20 330iA, J29 SPX30iA, G12 750Ld xDrive, G14 M850iA xDrive, F60 Countryman and G30 518d) show mainly an increased difference in the velocity range $< 90\,^{\mathrm{km}}/_{\mathrm{h}}$ compared to the road load determined with the coastdown method. Furthermore, the energy demand shares, which are transformed depending on the three different velocity ranges 'low' ($0\,^{\mathrm{km}}/_{\mathrm{h}} < v \leq 50\,^{\mathrm{km}}/_{\mathrm{h}}$), 'medium' ($50\,^{\mathrm{km}}/_{\mathrm{h}} < v \leq 90\,^{\mathrm{km}}/_{\mathrm{h}}$) and 'high' ($v > 90\,^{\mathrm{km}}/_{\mathrm{h}}$), are equivalent for these vehicles to the five vehicles investigated in subsection 4.2.1 (see Figure 4.24). The shares for the verification vehicles are illustrated in Figure C.2. This confirms the results, showed in subsection 4.2.3, that the road load in the velocity range $< 90\,^{\mathrm{km}}/_{\mathrm{h}}$ has a higher influence on the energy demand of the vehicle calculated on the basis of the WLTC than calculated on the basis of the ARTEMIS European driving cycle.

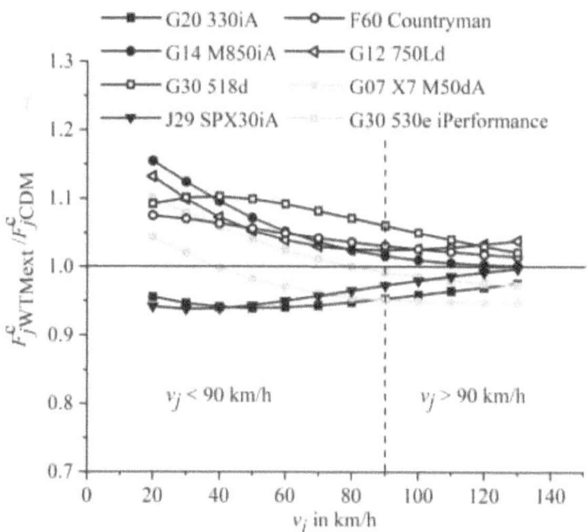

Figure C.1: Ratios of the road load determined with the wind tunnel method extended $F^c_{j,\mathrm{WTMext}}$ related to the coastdown method $F^c_{j,\mathrm{CDM}}$ over the reference velocity points v_j for the eight vehicles G20 330iA, G30 530e iPerformance, G07 X7 M50dA, J29 SPX30iA, G12 750Ld, G14 M850iA, F60 Cooper and G30 518d.

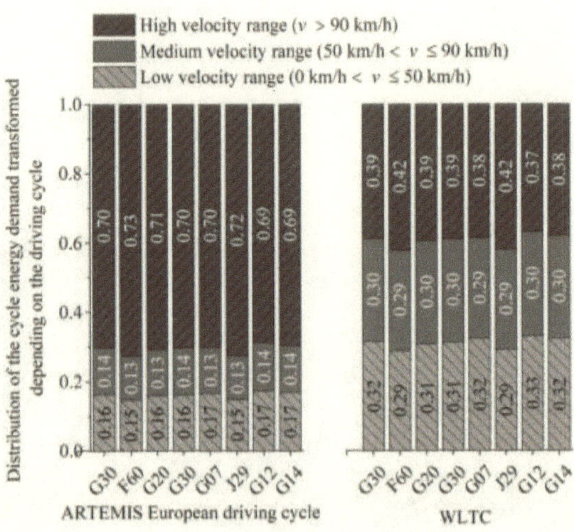

Figure C.2: Distribution of the energy demand transformed depending on the velocity ranges 'low', 'medium' and 'high' for the eight vehicles of the method verification and for both driving cycles.

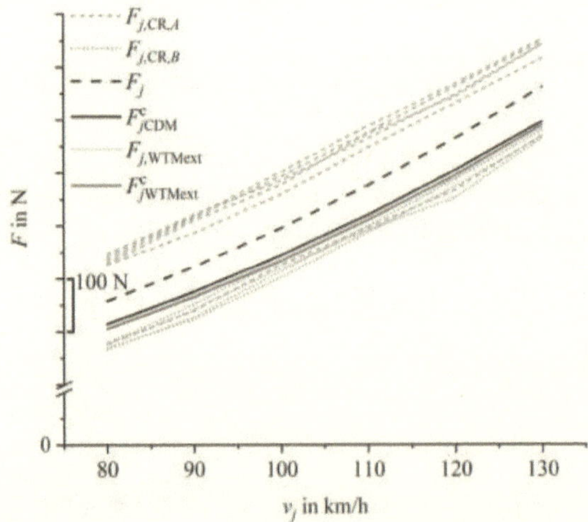

Figure C.3: Influence of the gear shifting of the 8HP automatic transmission on the road load determined using the coastdown method or the wind tunnel method extended in the velocity range from $80\,\mathrm{km/h}$ to $130\,\mathrm{km/h}$.

APPENDIX D

AEROLAB method

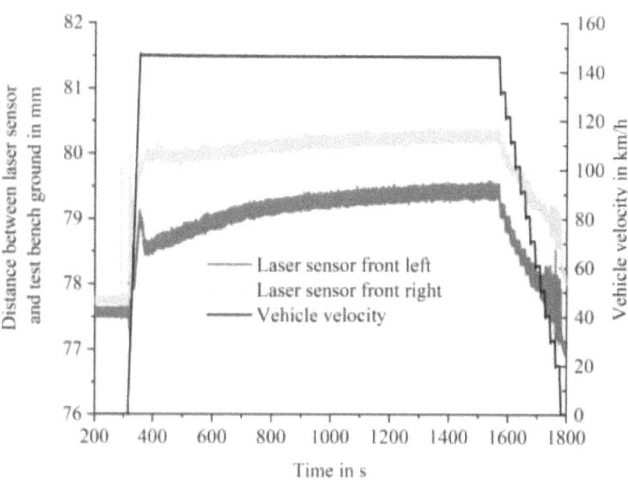

Figure D.1: Distance between the laser front left (dark grey line) and of the laser front right (grey line) mounted at the vehicle and the ground of the test bench as well as the vehicle velocity (black line) during the test procedure SV ALT woB at the flat belt dynamometer.

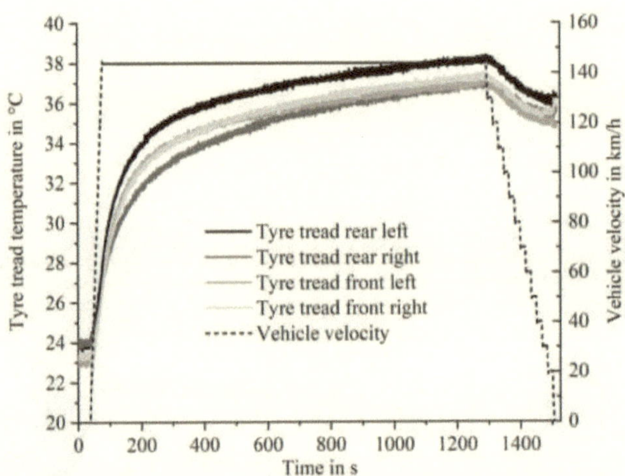

Figure D.2: Tyre tread temperature of the tyre front left (black line), front right (dark grey line), rear left (grey line) and rear right (light grey line) plotted over the test procedure SV ALT woB with the vehicle velocity (black dashed line).

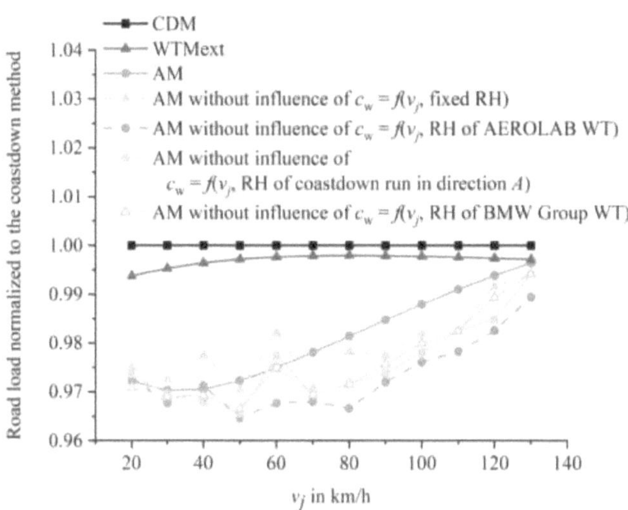

Figure D.3: Differences in the determined road load of a vehicle between the wind tunnel method extended (grey line with triangles), the AEROLAB method (grey line with dots) and the AEROLAB method without the influence of the aerodynamic drag depending on the wind velocity but fixed ride height (light grey dashed line with triangles), without the influence of the aerodynamic drag depending on the wind velocity and the ride height change from the AEROLAB wind tunnel (grey dashed line with dots), without the influence of the aerodynamic drag depending on the wind velocity and the ride height change of the coastdown runs in direction A (light grey line with squares) and without the influence of the aerodynamic drag depending on the wind velocity and the ride height change of BMW Group wind tunnel (light grey line with open triangles) normalized to the coastdown method (black line with squares).

In Figure D.4 the difference in the road load of the F60 Countryman measured using the AEROLAB method (grey line with open circles) and the road load of the F60 Countryman measured using the AEROLAB method but without the residual brake forces (grey dashed line with open circles) is illustrated.

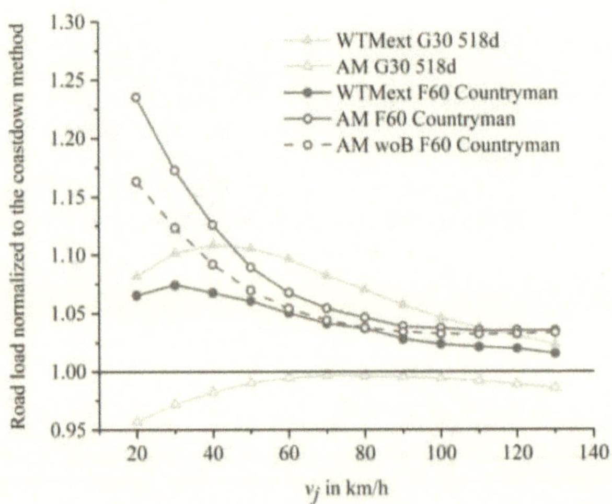

Figure D.4: Road load for the BMW G30 518d (light grey lines with triangles), the F60 Countryman (grey lines with dots) and the F60 Countryman without residual brake forces (grey dashed lines with dots) determined with the AEROLAB method (open symbols) and with the wind tunnel method extended (filled symbols) normalized with the corresponding road load determined with the coastdown method plotted over the reference velocity points v_j.

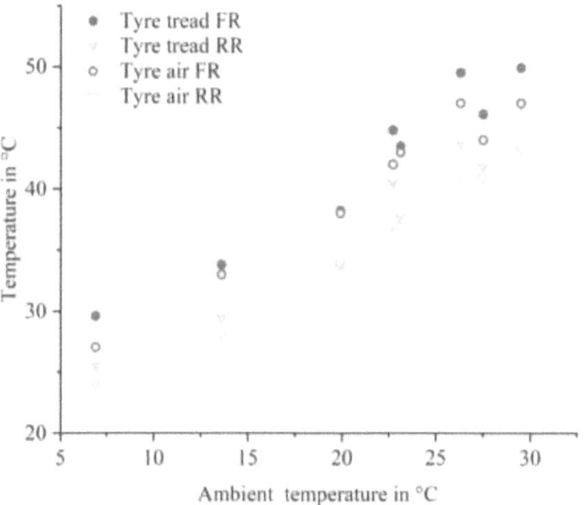

Figure D.5: Tyre air temperature (open symbols) measured with the Tire Pressure Monitoring System and the tyre tread temperatures (filled symbols) measured with infrared sensors installed in the wheel housings for the tyres front right FR (black symbols) and rear right RR (grey symbols) plotted over the ambient temperature for the test vehicle F46 216i.

APPENDIX E

Error calculation

In the following, the sensitivity coefficients c_m, which are used for the error calculation, are given.

E.1 Coastdown method - after polynomial regression

The equations for the sensitivity coefficients c_m of the error calculation after the polynomial regression:

$$c_{f_0} = \frac{\partial F_{j,\text{CDM}}}{\partial f_0} = \frac{2m_{\text{test}}}{m_1 + m_2} \left(K_0 \left(T_{\text{Amb}} - T_0 \right) + 1 \right) \tag{E.1}$$

$$c_{f_1} = \frac{\partial F_{j,\text{CDM}}}{\partial f_1} = v_j \left(K_0 \left(T_{\text{Amb}} - T_0 \right) + 1 \right) \tag{E.2}$$

$$c_{f_2} = -\frac{\partial F_{j,\text{CDM}}}{\partial f_2} = v_{\text{Wind}}^2 \left(K_0 \left(T_{\text{Amb}} - T_0 \right) + 1 \right) + \frac{100 \text{ kPa}}{p_{\text{Amb}}} \cdot \frac{T_{\text{Amb}}}{293 \text{ K}} \cdot v_j^2 \tag{E.3}$$

$$c_{m_1} = \frac{\partial F_{j,\text{CDM}}}{\partial m_1} = -2f_0 \cdot \frac{m_{\text{test}}}{\left(m_1 + m_2 \right)^2} \left(K_0 \left(T_{\text{Amb}} - T_0 \right) + 1 \right) \tag{E.4}$$

$$c_{m_2} = \frac{\partial F_{j,\text{CDM}}}{\partial m_2} = -2f_0 \frac{m_{\text{test}}}{\left(m_1 + m_2 \right)^2} \left(K_0 \left(T_{\text{Amb}} - T_0 \right) + 1 \right) \tag{E.5}$$

$$c_{T_{\text{Amb}}} = \frac{\partial F_{j,\text{CDM}}}{\partial T_{\text{Amb}}} = K_0 \left(-f_0 \left(1 - \frac{2m_{\text{test}}}{m_1 + m_2} \right) + f_0 + f_1 v_j - f_2 \cdot v_{\text{Wind}}^2 \right)$$
$$+ \frac{100 \text{ kPa}}{p_{\text{Amb}}} \cdot \frac{1}{293 \text{ K}} \cdot f_2 v_j^2 \quad \text{(E.6)}$$

$$c_{p_{\text{Amb}}} = \frac{\partial F_{j,\text{CDM}}}{\partial p_{\text{Amb}}} = -f_2 \cdot \frac{100 \text{ kPa}}{p_{\text{Amb}}^2} \cdot \frac{T_{\text{Amb}}}{293 \text{ K}} \cdot v_j^2 \quad \text{(E.7)}$$

$$c_{v_{\text{Wind}}} = \frac{\partial F_{j,\text{CDM}}}{\partial v_{\text{Wind}}} = -2 \cdot f_2 \cdot v_{\text{Wind}} \left(K_0 \left(T_{\text{Amb}} - T_0 \right) + 1 \right) \quad \text{(E.8)}$$

Table E.1: Modified components of the combined standard uncertainty for the coastdown method (after the polynomial regression)

v_j in $^{\text{km}}/_{\text{h}}$	20	30	40	50	60	70	80	90	100	110	120	130
$u^*_{f_0, F_{j,\text{CDM}}}$ in N	6.0	6.0	6.0	6.0	6.0	6.0	6.0	6.0	6.0	6.0	6.0	6.0
$u^*_{f_1, F_{j,\text{CDM}}}$ in N	6.0	9.0	12.0	14.9	17.9	20.9	23.9	26.9	29.9	32.9	35.9	38.8
$u^*_{f_2, F_{j,\text{CDM}}}$ in N	1.8	4.1	7.4	11.6	16.7	22.8	29.7	37.6	46.5	56.3	67.0	78.6
$u^*_{T_{\text{Amb}}, F_{j,\text{CDM}}}$ in N	0.1	0.1	0.1	0.1	0.1	0.1	0.1	0.1	0.1	0.1	0.2	0.2
$u^*_{p_{\text{Amb}}, F_{j,\text{CDM}}}$ in N	0.0	0.0	0.0	0.0	0.0	0.0	0.0	0.0	0.0	0.0	0.1	0.1
$u^*_{v_{\text{Wind}}, F_{j,\text{CDM}}}$ in N	0.0	0.0	0.0	0.0	0.0	0.0	0.0	0.0	0.0	0.0	0.0	0.0
$u^*_{F_{\text{corr}}, F_{j,\text{CDM}}}$ in N	4.6	3.4	3.0	2.9	3.0	3.0	2.8	2.8	3.3	4.5	6.3	8.7
$u^*_{m, F_{j,\text{CDM}}}$ in N	0.4	0.4	0.4	0.4	0.4	0.4	0.4	0.4	0.4	0.4	0.4	0.4

E.2 Wind tunnel method extended

E.2.1 Flat belt dynamometer part

In the following, the equations for the sensitivity coefficients c_m for the flat belt dynamometer part are stated:

$$c_{F_{j,\text{Dyno}}} = \frac{\partial F_{j,\text{WTMext}}}{\partial F_{j,\text{Dyno}}} = 2 \cdot K_{\text{ext}} \cdot \frac{m_{\text{test}}}{m_1 + m_2} \left(K_0 \left(T_{\text{Dyno}} - T_0 \right) + 1 \right) \quad \text{(E.9)}$$

$$c_{T_{\text{Dyno}}} = \frac{\partial F_{j,\text{WTMext}}}{\partial T_{\text{Dyno}}} = K_0 \cdot K_{\text{ext}} \left(F_{j,\text{Dyno}} - F_{j,\text{Dyno}} \left(-2 \cdot \frac{m_{\text{test}}}{m_1 + m_2} + 1 \right) \right) \tag{E.10}$$

$$c_{F_{j,\text{Drive}}} = \frac{\partial F_{j,\text{WTMext}}}{\partial F_{j,\text{Drive}}} = 1 - K_{\text{ext}} \tag{E.11}$$

$$c_{m_1} = \frac{\partial F_{j,\text{WTMext}}}{\partial m_1} = -2 \cdot F_{j,\text{Dyno}} \cdot K_{\text{ext}} \frac{m_{\text{test}}}{(m_1 + m_2)^2} \left(K_0 \left(T_{\text{Dyno}} - T_0 \right) + 1 \right) \tag{E.12}$$

$$c_{m_2} = \frac{\partial F_{j,\text{WTMext}}}{\partial m_2} = -2 \cdot F_{j,\text{Dyno}} \cdot K_{\text{ext}} \frac{m_{\text{test}}}{(m_1 + m_2)^2} \left(K_0 \left(T_{\text{Dyno}} - T_0 \right) + 1 \right) \tag{E.13}$$

Table E.2: Modified components of the combined standard uncertainty for the flat belt dynamometer part of the wind tunnel method extended

v_j in $^{\text{km}}/_{\text{h}}$	20	30	40	50	60	70	80	90	100	110	120	130
$u^*_{F,F_j,\textbf{WTMext}}$ in N	2.8	2.8	2.8	2.8	2.8	2.8	2.8	2.8	2.8	2.8	2.8	2.8
$u^*_{F_j,\textbf{Dyno},F_j,\textbf{WTMext}}$ in N	12.7	14.3	16.6	16.2	17.2	15.4	17.0	16.9	20.3	20.3	24.5	24.7
$u^*_{p_{\text{Tyre}},F_j,\textbf{WTMext}}$ in N	3.6	3.6	3.6	3.6	3.6	3.6	3.6	3.6	3.6	3.6	3.6	3.6
$u^*_{F_{\text{fluc}},F_j,\textbf{WTMext}}$ in N	0.1	0.1	0.0	0.0	0.0	0.0	0.0	0.0	0.1	0.1	0.3	0.1
$u^*_{\text{TOM},F_j,\textbf{WTMext}}$ in N	0.5	0.5	0.5	0.5	0.5	0.5	0.5	0.5	0.5	0.5	0.5	0.5
$u^*_{T_{\text{dyno}},F_j,\textbf{WTMext}}$ in N	0.5	0.5	0.6	0.6	0.6	0.6	0.6	0.6	0.6	0.6	0.7	0.7
$u^*_{m,F_j,\textbf{WTMext}}$ in N	1.4	1.4	1.5	1.5	1.5	1.6	1.6	1.6	1.7	1.7	1.7	1.8

Table E.3: Standard deviation of the velocity of the flat belts at the flat belt dynamometer

v_j in $^{\text{km}}/_{\text{h}}$	20	30	40	50	60	70	80	90	100	110	120	130
$\sigma\left(v_j\right)$ in $^{\text{km}}/_{\text{h}}$	0.01	0.01	0.01	0.01	0.01	0.01	0.01	0.01	0.01	0.01	0.01	0.01

E.2.2 Wind tunnel part

In the following, the equations for the sensitivity coefficients c_m for the wind tunnel part are stated:

$$c_{F_{x,\text{Wind}}} = \frac{\partial F_{j,\text{WTMext}}}{\partial F_{x,\text{Wind}}} = \frac{\rho}{2} \cdot \frac{\kappa - 1}{\kappa} \cdot \frac{v_j^2}{p_{\text{stat}} \left(\left(\frac{\Delta p}{p_{\text{stat}}} + 1 \right)^{\frac{\kappa-1}{\kappa}} - 1 \right)} \tag{E.14}$$

$$c_{p_{\text{stat}}} = \frac{\partial F_{j,\text{WTMext}}}{\partial p_{\text{stat}}} = \frac{\rho}{2} \cdot \frac{\Delta p}{\left(\frac{\kappa}{\kappa-1} \right)^2 p_{\text{stat}}^3 \left(\frac{\Delta p}{p_{\text{stat}}} + 1 \right)^{1-\frac{\kappa-1}{\kappa}} \left(\left(\frac{\Delta p}{p_{\text{stat}}} + 1 \right)^{\frac{\kappa-1}{\kappa}} - 1 \right)^2} \cdot v_j^2$$
$$- \frac{\rho}{2} \cdot \frac{\kappa - 1}{\kappa} \cdot \frac{F_{x,\text{Wind}}}{p_{\text{stat}}^2 \left(\left(\frac{\Delta p}{p_{\text{stat}}} + 1 \right)^{\frac{\kappa-1}{\kappa}} - 1 \right)} \cdot v_j^2 \tag{E.15}$$

$$c_{\Delta p} = \frac{\partial F_{j,\text{WTMext}}}{\partial \Delta p} = -\frac{\rho}{2} \cdot \frac{1}{\left(\frac{\kappa}{\kappa-1} \right)^2 p_{\text{stat}}^2 \left(\frac{\Delta p}{p_{\text{stat}}} + 1 \right)^{1-\frac{\kappa-1}{\kappa}} \left(\left(\frac{\Delta p}{p_{\text{stat}}} + 1 \right)^{\frac{\kappa-1}{\kappa}} - 1 \right)^2} \cdot v_j^2 \tag{E.16}$$

Table E.4: Modified components of the combined standard uncertainty for the wind tunnel part of the wind tunnel method extended

v_j in $^{\text{km}}/_{\text{h}}$	20	30	40	50	60	70	80	90	100	110	120	130
$u^*_{F_{x,\text{acc}},F_{j,\text{WTMext}}}$ in N	0.0	0.1	0.1	0.2	0.3	0.4	0.6	0.7	0.9	1.1	1.3	1.5
$u^*_{F_{x,\text{res}},F_{j,\text{WTMext}}}$ in N	0.0	0.0	0.1	0.1	0.1	0.2	0.2	0.3	0.4	0.4	0.5	0.6
$u^*_{F_{x,\text{fluc}},F_{j,\text{WTMext}}}$ in N	0.0	0.0	0.0	0.0	0.0	0.1	0.1	0.1	0.1	0.1	0.2	0.2
$u^*_{p_\text{s},F_{j,\text{WTMext}}}$ in N	0.0	0.0	0.0	0.0	0.0	0.0	0.0	0.0	0.0	0.0	0.0	0.0
$u^*_{p_\text{s,fluc},F_{j,\text{WTMext}}}$ in N	0.0	0.0	0.0	0.0	0.0	0.0	0.0	0.0	0.0	0.0	0.0	0.0
$u^*_{\Delta p,F_{j,\text{WTMext}}}$ in N	0.0	0.0	0.0	0.0	0.0	0.0	0.0	0.0	0.0	0.0	0.0	0.0
$u^*_{\Delta p_\text{fluc},F_{j,\text{WTMext}}}$ in N	0.0	0.0	0.0	0.0	0.0	0.0	0.0	0.0	0.0	0.0	0.0	0.0

E.3 AEROLAB method

In the following, the equations for the sensitivity coefficients c_m for the AEROLAB method are given:

$$c_{f_0} = \frac{\partial F_{j,\text{AM}}}{\partial f_0} = 2 \frac{m_{\text{test}}}{m_1 + m_2} \left(K_0 \left(T_{\text{Amb}} - T_0 \right) + 1 \right) \tag{E.17}$$

$$c_{f_1} = \frac{\partial F_{j,\mathrm{AM}}}{\partial f_1} = \left(K_0\left(T_{\mathrm{Amb}} - T_0\right) + 1\right) v_j \tag{E.18}$$

$$c_{f_2} = \frac{\partial F_{j,\mathrm{AM}}}{\partial f_2} = \frac{100\ \mathrm{kPa}}{p_{\mathrm{Amb}}} \cdot \frac{T_{\mathrm{Amb}}}{293\ \mathrm{K}} \cdot v_{Veh}^2 \tag{E.19}$$

$$c_{m_1} = \frac{\partial F_{j,\mathrm{AM}}}{\partial m_1} = -2 f_0 \frac{m_{\mathrm{test}}}{\left(m_1 + m_2\right)^2} \left(K_0\left(T_{\mathrm{Amb}} - T_0\right) + 1\right) \tag{E.20}$$

$$c_{m_2} = \frac{\partial F_{j,\mathrm{AM}}}{\partial m_2} = -2 f_0 \frac{m_{\mathrm{test}}}{\left(m_1 + m_2\right)^2} \left(K_0\left(T_{\mathrm{Amb}} - T_0\right) + 1\right) \tag{E.21}$$

$$c_{T_{\mathrm{Amb}}} = \frac{\partial F_{j,\mathrm{AM}}}{\partial T_{\mathrm{Amb}}} = K_0 \left(f_0 - f_0 \left(1 - 2\frac{m_{\mathrm{test}}}{m_1 + m_2}\right) + f_1 v_j\right) + \frac{100\ \mathrm{kPa}}{p_{\mathrm{Amb}}} \cdot \frac{1}{293\ \mathrm{K}} f_2 v_j^2 \tag{E.22}$$

$$c_{p_{\mathrm{Amb}}} = \frac{\partial F_{j,\mathrm{AM}}}{\partial p_{\mathrm{Amb}}} = -\frac{100\ \mathrm{kPa}}{p_{\mathrm{Amb}}^2} \cdot \frac{T_{\mathrm{Amb}}}{293\ \mathrm{K}} f_2 v_j^2 \tag{E.23}$$

$$c_{F_{\mathrm{Brake}}} = \frac{\partial F_{j,\mathrm{AM}}}{\partial F_{\mathrm{Brake}}} = 1 \tag{E.24}$$

$$c_{F_{j,\mathrm{WT,woB}}} = \frac{\partial F_{j,\mathrm{AM}}}{\partial F_{j,\mathrm{WT,woB}}} = 1 \tag{E.25}$$

Table E.5: Modified components of the combined standard uncertainty for the AERO-LAB method

v_j in $\mathrm{km/h}$	20	30	40	50	60	70	80	90	100	110	120	130
$u^*_{f_0,F_j,\mathrm{AM}}$ in N	2.1	2.1	2.1	2.1	2.1	2.1	2.1	2.1	2.1	2.1	2.1	2.1
$u^*_{F_{\mathrm{fluc}},F_j,\mathrm{AM}}$ in N	0.2	0.1	0.1	0.2	0.2	0.2	0.3	0.4	0.3	0.6	0.4	0.6
$u^*_{f_1,F_j,\mathrm{AM}}$ in N	0.0	0.0	0.0	0.0	0.0	0.0	0.0	0.0	0.0	0.0	0.0	0.0
$u^*_{f_2,F_j,\mathrm{AM}}$ in N	0.0	0.0	0.0	0.0	0.0	0.0	0.0	0.0	0.0	0.0	0.0	0.0
$u^*_{T_{\mathrm{Amb,acc}},F_j,\mathrm{AM}}$ in N	0.5	0.5	0.6	0.6	0.7	0.7	0.8	0.9	1.0	1.1	1.2	1.3
$u^*_{T_{\mathrm{Amb,fluc}},F_j,\mathrm{AM}}$ in N	0.4	0.4	0.5	0.5	0.6	0.6	0.7	0.7	0.8	0.9	1.0	1.1
$u^*_{p_{\mathrm{Amb,acc}},F_j,\mathrm{AM}}$ in N	0.0	0.0	0.0	0.0	0.0	0.0	0.0	0.0	0.0	0.0	0.0	0.0
$u^*_{p_{\mathrm{Amb,fluc}},F_j,\mathrm{AM}}$ in N	0.0	0.1	0.2	0.3	0.4	0.6	0.7	0.9	1.2	1.4	1.7	2.0
$u^*_{F_{\mathrm{Brake}},F_j,\mathrm{AM}}$ in N	6.5	6.2	6.2	6.7	6.5	6.4	6.3	6.1	6.2	6.3	6.1	6.4
$u^*_{F_{j,\mathrm{WT,woB}},F_j,\mathrm{AM}}$ in N	4.1	5.2	6.1	6.7	7.0	7.0	6.7	6.1	5.4	4.8	4.8	5.9
$u^*_{m,F_j,\mathrm{AM}}$ in N	1.0	1.0	1.0	1.0	1.0	1.0	1.0	1.0	1.0	1.0	1.0	1.0

List of Figures

List of Tables

Abbreviations

Abbreviation	Meaning
ALT	ALTernative warm-up phase
AM	AEROLAB Method
ARTEMIS	Assessment and Reliability of Transport Emission Models and Inventory Systems
AT	Automatic Transmission
AWD	All-Wheel Drive
abbr.	Abbreviation
BMW	Bayerische Motoren Werke
CD	CoastDown (= measurement procedure by deceleration)
CD ALT wB	Test procedure according to WLTP with braking phase (wB), alternative warm-up phase (ALT) and measurement procedure by deceleration (CD)
CD ALT woB	Test procedure without braking phase (woB), alternative warm-up phase (ALT) and measurement procedure by deceleration (CD)
CD STD wB	Test procedure according to WLTP with braking phase (wB), standard warm-up phase (STD) and measurement procedure by deceleration (CD)
CD STD woB	Test procedure without braking phase (woB), standard warm-up phase (STD) and measurement procedure by deceleration (CD)
CDM	CoastDown Method
CR	Coastdown Run
DIN	Deutsches Institut für Normung
DRIVE	Dedicated Road Infrastructure for Vehicle safety in Europe
Dyno	Dynamometer
EN	Europäische Norm

Continued on next page

Abbreviation	Meaning
EU	European Union
FA	Front Axle
FL	Front Left
FR	Front Right
FKFS	Forschungsinstitut für Kraftfahrwesen und Fahrzeugmotoren Stuttgart
FWD	Front Wheel Drive
GRG	Generalized Reduced Gradient
GPS	Global Positioning System
GTR	Global Technical Regulations
GUM	Guide to the expression of Uncertainty in Measurement
HFMS	Horizontal Force Measurement System
HYZEM	European development of HYbrid vehicle technology approaching efficient Zero EMission mobility
ISO	International Organisation for Standardization
LC	Load Cell
Meas.	Measurement
MIRO	Mass In Running Order
MODEM	MODelling of EMissions and fuel consumption in urban areas
MT	Manual Transmission
NEDC	New European Driving Cycle
PDF	Probability Density Function
PDT	Difference Pressure Transducer
PT	Pressure Transducer
RA	Rear Axle
RH	Ride Height
RL	Rear Left
RR	Rear Right
RWD	Rear Wheel Drive
STD	STanDard warm-up phase
SV	Stabilized Velocity
SV ALT wB	Test procedure according to WLTP with braking phase (wB), alternative warm-up phase (ALT) and measurement procedure with stabilized velocity (SV)
SV ALT woB	Test procedure without braking phase (woB), alternative warm-up phase (ALT) and measurement procedure with stabilized velocity (SV)
SV STD wB	Test procedure according to WLTP with braking phase (wB), standard warm-up phase (STD) and measurement procedure with stabilized velocity (SV)

Continued on next page

Abbreviation	Meaning
SV STD woB	Test procedure without braking phase (woB), standard warm-up phase (STD) and measurement procedure with stabilized velocity (SV)
std	Standard
TOM	TOrque Meter
TPMS	Tire Pressure Monitoring System
TS	Test Section
UN-ECE	United Nations Economic Comission for Europe
USA	United States of America
Veh	Vehicle
ViL	Vehicle-in-the-Loop test bench
WDU	Wheel Drive Unit
WLTC	Worldwide Light-duty driving Test Cycle
WLTP	Worldwide harmonized Light vehicles Test Procedures
WT	Wind Tunnel
WTM	Wind Tunnel Method
WTMext	Wind Tunnel Method extended
wB	With Braking phase
woB	WithOut Braking phase

Symbols

Latin symbols

Symbols	Unit	Meaning
A		Obligatory driving direction at a test track
A_x	m^2	Frontal area of a vehicle and of an individual vehicle in a road load matrix family
$A_{x,r}$	m^2	Frontal area of the representative vehicle in a road load matrix family
a	m/s^2	Translational acceleration
a		Constant of an equation
a_d	N	First dynamometer set coefficient
a_i	m/s^2	Acceleration of the vehicle during the time period $(i-1)$ to i at the time step i
a_j	m/s^2	Deceleration at reference velocity point v_j
a_-, a_+		Interval boundaries
B		Obligatory driving direction at a test track
b		Constant of an equation

Continued on next page

Symbols	Unit	Meaning
b_d	$\mathrm{N}/(\mathrm{m/s})$	Second dynamometer set coefficient
C_j	Nm	Running resistance measured with the torque meter method according to WLTP
C_α	$\mathrm{N}/\!^\circ$	Cornering stiffness of the tyre
C_γ	$\mathrm{N}/\!^\circ$	Camber stiffness of the tyre
c		Constant of an equation
c_A		Lift/Downward force coefficient
c_d	$\mathrm{N}/(\mathrm{m/s})^2$	Dynamometer set coefficient
c_m, c_l		Sensitivity coefficient of the expectation values x_m amd x_l of the input quantities
c_T		Tangential force coefficient
c_Vent		Aerodynamic drag coefficient of the rotating tyre/Wheel ventilation resistance coefficient
c_w		Aerodynamic drag coefficient
$c_\mathrm{w(140,fixed)}$		Aerodynamic drag coefficient measured in the BMW Group wind tunnel at $140\,\mathrm{km/h}$ and fixed ride height
$c_\mathrm{w(}v_j\mathrm{,float)}$		Aerodynamic drag coefficient measured in the BMW Group wind tunnel with the ride height depending on the wind velocity
c_w^*		Extended aerodynamic drag coefficient
d	m	Distance
d_i	m	Distance the vehicle is travelling during the time period $(i-1)$ to i
d_FA	mm	Distance of the outer edge of the front wheel to the reference edge
d_RA	mm	Distance of the outer edge of the rear wheel to the reference edge
E	J	Total energy demand for a vehicle for the chosen test cycle
E_AM	J	Total energy demand over a complete test cycle calculated with the road load determined using the AEROLAB Method (AM)
E_CDM	J	Total energy demand over a complete test cycle calculated with the road load determined using the CoastDown Method (CDM)
E_WTM	J	Total energy demand over a complete test cycle calculated with the road load determined using the Wind Tunnel Method (WTM)
E_WTMext	J	Total energy demand over a complete test cycle calculated with the road load determined using the Wind Tunnel Method extended (WTMext)

Continued on next page

Symbols	Unit	Meaning
E_i	J	Energy demand of a vehicle during a specific time period $(i-1)$ to i
e		Exponent value
e_{v_j}		Exponent value describing the influence of the tyre inflation pressure on the rolling resistance $F_{j,\text{Roll}}$ at the reference velocity point v_j and the investigated tyre inflation pressure p_{Tyre}
e_{mass}		Mass factor
e_0	m	Distance between the centre of the wheel and the resulting force of the pressure surface of the tyre contact area
F	N	Force
F_{Air}	N	Aerodynamic drag
F_{Climb}	N	Climbing resistance
F_{Drive}	N	Drivetrain losses
F_{Inertial}	N	Inertial resistance
F_i	N	Force for driving the vehicle during the time period $(i-1)$ to i
F_j	N	Uncorrected road load of the vehicle at the reference velocity point v_j
$F_{j,\text{A}}$	N	Lift/downward forces at the reference velocity point v_j
$F_{j,\text{Air}}$	N	Aerodynamic drag at the reference velocity point v_j
$F_{j,\text{AM}}$	N	Road load of the vehicle corrected to reference conditions at the reference velocity point v_j using the AEROLAB method
$F_{j,\text{AM}}^c$	N	Road load of a vehicle at the reference velocity point v_j for the AEROLAB method calculated on the basis of the road load coefficients f_0^c, f_1^c and f_2^c
$F_{j,\text{Brake}}$	N	Residual brake force at the reference velocity point v_j
$F_{j,\text{Calc}}$	N	Calculated road load force at the reference velocity point v_j
$F_{j,\text{CDM}}$	N	Road load of a vehicle at the reference velocity point v_j for the coastdown method calculated on the basis of the road load coefficients f_0, f_1 and f_2
$F_{j,\text{CDM}}^c$	N	Road load of the vehicle corrected to reference conditions at the reference velocity point v_j according to the coastdown method calculated on the basis of the road load coeffcients f_0^c, f_1^c and f_2^c
$F_{j,\text{CR},A}$	N	Calculated road load for one Coastdown Run (CR) in direction A using the coastdown run time t_{jAi}
$F_{j,\text{CR},B}$	N	Calculated road load for one Coastdown Run (CR) in direction B using the coastdown run time t_{jBi}
$F_{j,\text{c}}$	N	Calculated road load force for an individual vehicle in a road load matrix family at the reference velocity point v_j

Continued on next page

Symbols	Unit	Meaning
$F_{j,\mathrm{cd}}$	N	Calculated road load force based on the vehicle parameters test mass, vehicle height and width at the reference velocity point v_j
$F_{j,\mathrm{Drive}}$	N	Drivetrain losses of the vehicle at the reference velocity point v_j and part of the vehicle road load measured at the flat belt dynamometer $F_{j,\mathrm{Dyno}}$
$F_{j,\mathrm{Drive,Wheel}}$	N	Wheel specific drivetrain losses at the reference velocity v_j
$F_{j,\mathrm{Dyno}}$	N	Corrected road load of a vehicle at the reference velocity point v_j to reference conditions measured at the flat belt dynamometer
$F_{j,\mathrm{Dyno,woB}}$	N	Road load of a vehicle without residual brake forces measured at the flat belt dynamometer
$F^*_{j,\mathrm{Dyno}}$	N	Uncorrected road load of the vehicle at the reference velocity point v_j determined using the flat belt dynamometer
$F_{j,\mathrm{d}}$	N	Dynamometer load setting
$F_{j,\mathrm{Roll}}$	N	Rolling resistance of a vehicle at the reference velocity point v_j and part of the vehicle road load measured at the flat belt dynamometer $F_{j,\mathrm{Dyno}}$
$F_{\mathrm{j,Roll,Lift}}$	N	Rolling resistance reduced by lift forces
$F_{j,\mathrm{TOM,Wheel}}$	N	Wheel specific force determined with TOM at the reference velocity v_j
$F_{j,\mathrm{WT,woB}}$	N	Road load of a vehicle without residual brake forces measured in the AEROLAB wind tunnel
$F_{j,\mathrm{WTM}}$	N	Road load of a vehicle according to the wind tunnel method
$F^{\mathrm{c}}_{j,\mathrm{WTM}}$	N	Corrected road load of a vehicle according to the wind tunnel method calculated on the basis of the road load coefficients f^c_0, f^c_1 and f^c_2
$F_{j,\mathrm{WTMext}}$	N	Road load of a vehicle according to the wind tunnel method extended with the additional correction factor K_{ext}
$F^{\mathrm{c}}_{j,\mathrm{WTMext}}$	N	Corrected road load of a vehicle according to the wind tunnel method extended with the additional correction factor K_{ext} calculated on the basis of the road load coefficients f^c_0, f^c_1 and f^c_2
$F_{j,x}$	N	Force in x direction at the reference velocity v_j
$F_{\mathrm{LC,FA}}$	N	Sum of the measured forces at the both load cells (LC) of the front WDUs
$F_{\mathrm{LC,RA}}$	N	Sum of the measured forces at the both load cells (LC) of the rear WDUs
F_{lat}	N	Lateral tyre force
F_{N}	N	Wheel load
$F_{\mathrm{N,ISO}}$	N	Wheel load as specified in ISO 8767

Continued on next page

Symbols	Unit	Meaning
F_{Roll}	N	Rolling resistance
$F_{\text{Roll},\alpha}$	N	Resistance due to tyre slip angle α
$F_{\text{Roll,Brake}}$	N	Residual brake force
$F_{\text{Roll,Camber}}$	N	Resistance due to the tyre camber angle γ
$F_{\text{Roll,Fric}}$	N	Losses due to bearing friction and residual brake forces
$F_{\text{Roll,ISO}}$	N	Rolling resistance measured according to ISO 9948
$F_{\text{Roll,Ref}}$	N	Rolling resistance adjusted to reference temperature T_0
$F_{\text{Roll,Road}}$	N	Road rolling resistance
$F_{\text{Roll,Toe}}$	N	Resistance due to the tyre toe angle δ_0
$F_{\text{Roll,Tyre}}$	N	Tyre rolling resistance
$F_{\text{Roll,Tyre,Air}}$	N	Aerodynamic drag of the rotating tyre
$F_{\text{Roll,Tyre,Flex}}$	N	Flexing resistance
$F_{\text{Roll,Tyre,Fric}}$	N	Frictional resistance
$F_{\text{Roll,10,uncorr}}$	N	Rolling resistance measured at the flat belt dynamometer at an ambient temperature of $10\,°\text{C}$
$F_{\text{Roll,23,corr}}$	N	Rolling resistance measured at the flat belt dynamometer at an ambient temperature of $23\,°\text{C}$, but corrected to the reference temperature of $20\,°\text{C}$
F_{Rot}	N	Rotational force
F_{T}	N	Tangential force
F_{Trans}	N	Translational force
$F_{\text{Veh,FA}}$	N	Vehicle losses at the front axle of a vehicle
$F_{\text{Veh,RA}}$	N	Vehicles losses at the rear axle of a vehicle
F_{Vent}	N	Wheel ventilation resistance, which is equal to the aerodynamic drag of the rotating tyre $F_{\text{Roll,Tyre,Air}}$
$F_{x,v_{\text{high}}}$	N	Losses in x direction at the high velocity of $180\,\text{km/h}$
$F_{x,v_{\text{low}}}$	N	Losses in x direction at the low velocity of $40\,\text{km/h}$
$F_{x,\text{Wind}}$	N	Force in x direction of the aerodynamic drag
F_{Z}	N	Total vehicle weight force
f		Functional relationship of input quantities
f_0	N	Constant term of the road load coefficients
f_0^{c}	N	Constant term of the corrected road load coefficients
$f_{0,\text{r}}$	N	Constant term of the road load coefficients of the representative vehicle in the road load matrix family
f_1	$\text{N}/(\text{m/s})$	Coefficient of the first order term of the road load coefficients
f_1^{c}	$\text{N}/(\text{m/s})$	Coefficient of the first order term of the corrected road load coefficients
f_2	$\text{N}/(\text{m/s})^2$	Coefficient of the second order term of the road load coefficients
f_2^{c}	$\text{N}/(\text{m/s})^2$	Coefficient of the second order term of the corrected road load coefficients

Continued on next page

Symbols	Unit	Meaning
$f_{2,\mathrm{r}}$	$\mathrm{N}/_{(\mathrm{m/s})^2}$	Coefficient of the second order term of the road load coefficients of the representative vehicle in the road load matrix family
f_{r}		Rolling resistance coefficient
$f_{\mathrm{r,Safety\ Walk}}$		Rolling resistance coefficient on a Safety Walk^{TM} surface
$f_{\mathrm{r,Steel}}$		Rolling resistance coefficient on a steel surface
f_{r0}		Constant term of the rolling resistance coefficient f_{r}
f_{r1}		First order term of the rolling resistance coefficient f_{r}
f_{r4}		Fourth order term of the rolling resistance coefficient f_{r}
g_{Earth}	$\mathrm{m}/_{\mathrm{s}^2}$	Earth's gravity (set to 9.81 $\mathrm{m/s^2}$)
$g_{X_m}(\xi_m)$		Probability density function of the possible value ξ_m of the quantity X_m
h		Coefficient due to the Student t-distribution
h_{Veh}	m	Vehicle height as defined in Standard ISO 612:1978
i		Continuous variable
J_{red}	$\mathrm{kg} \cdot \mathrm{m}^2$	Reduced rotational inertia of the vehicle components
j		Counter variable for the reference velocity point v_j (20 $\mathrm{km/h}$, 30 $\mathrm{km/h}$, ..., 130 $\mathrm{km/h}$)
K_{ext}		Correction factor for the wind tunnel method extended
K_0	K^{-1}	Correction factor for the rolling resistance (and the drivetrain losses)
K_1		Test mass correction factor
K_2		Air resistance correction factor
k		Continuous variable
k_p		Coverage factor
k_{r}		Factor, which considers the inertial resistances of the drivetrain during acceleration and deceleration
l		Continuous variable
$l_{\mathrm{Track,FA}}$	mm	Track at the front axle
$l_{\mathrm{Track,RA}}$	mm	Track at the rear axle
l_{WB}	mm	Wheel base of the vehicle
$M_{j,\mathrm{TOM,Wheel}}$	Nm	Wheel specific torque measured with TOM at the reference velocity v_j
$M_{\mathrm{Roll,Tyre,Air}}$	Nm	Aerodynamic resistance moment
m		Continuous variable
m_{av}	kg	Arithmetic average of the test vehicle masses at the beginning and at the end of the test procedure
m_{eq}	kg	Equivalent vehicle mass
m_{MIRO}	kg	Sum of the mass in running order
m_{r}	kg	Equivalent effective mass of rotating components

Continued on next page

Symbols	Unit	Meaning
m_test	kg	Vehicle test mass defined for the road load determination and also for an individual vehicle in a road load matrix family
$m_\text{test,r}$	kg	Test mass of the representative vehicle in a road load matrix family
m_Veh	kg	Vehicle mass
m_1	kg	Test vehicle mass at the beginning of the coastdown test procedure
m_2	kg	Test vehicle mass at the end of the coastdown test procedure
N		Number of input quantities X_m
n		Number of pairs of coastdown runs or number of independent repeated observations $q_{m,k}$
P_Bearing	W	Bearing losses of a gear transmission
P_Comp	W	Losses of further aggregates and components of a gear transmission
P_Gear	W	Gearing losses of a gear transmission
P_L	W	Load dependent losses of the gear transmission
P_Seal	W	Seal losses of a gear transmission
$P_\text{Splashing}$	W	Splashing losses of a gear transmission
P_0	W	Load independent losses of the gear transmission
p_Amb	kPa	Atmospheric pressure
$\overline{p}_\text{Amb}$	kPa	Average atmospheric pressure
p_j		Statistical precision for the reference velocity point v_j
p_stat	bar	Static pressure
p_Tyre	bar	Tyre inflation pressure
$p_\text{Tyre,ISO}$	bar	Tyre inflation pressure as defined in ISO 8767
$p_\text{Tyre,Std}$	bar	Tyre inflation pressure according to manufacturer specification
p_0	kPa	Reference atmospheric pressure of $100\,\text{kPa}$
$\overline{q_m}$		Arithmetic mean or average of n independent repeated observations $q_{m,k}$
$q_{m,k}$		Independent observation
q_TS	Pa	Test section dynamic pressure
R	J/kg·K	Specific gas constant with the value of $287.0\ \text{J/kg·K}$
R_Tyre	kg/t	Tyre rolling resistance of an individual vehicle in a road load matrix family
$R_\text{Tyre,r}$	kg/t	Tyre rolling resistance of the representative vehicle in a road load matrix family
$r\left(x_m, x_l\right)$		Correlation coefficient between the expectation values x_m and x_l
r_dyn	m	Dynamic rolling radius
r_stat	m	Distance between wheel centre and road surface

Continued on next page

Symbols	Unit	Meaning
r^2		Coefficient of determination
s_T	m	Displacement of the tyre deformation
T_{Amb}	K	Ambient atmospheric temperature
$\overline{T}_{\mathrm{Amb}}$	K	Arithmetic average ambient atmospheric temperature
$\overline{T}_{\mathrm{Dyno}}$	K	Arithmetic average temperature at the test bench during the measurement procedure
T_0	K	Reference atmospheric temperature of $20\,^{\circ}\mathrm{C}$
t	s	Integration variable
t_{end}	s	Time, at which the chosen test cycle ends
t_i	s	Time step/elapsed time during the deceleration
t_{jAi}	s	Coastdown time of the $\mathrm{i^{th}}$ measurement at the reference velocity point v_j in direction A
t_{jBi}	s	Coastdown time of the $\mathrm{i^{th}}$ measurement at the reference velocity point v_j in direction B
t_{start}	s	Time, at which the chosen test cycle starts
t_0	s	Start time of the deceleration
U_{dyn}	mm	Dynamic circumference
U_{TS}	m/s	Test section air velocity
U_y		Expanded uncertainty
$u^*_{\Delta p, F_j, \mathrm{WTMext}}$	N	Modified component of the combined standard uncertainty for the accuracy of the measured differential pressure Δp using the wind tunnel method extended
$u^*_{\Delta p_{\mathrm{fluc}}, F_j, \mathrm{WTMext}}$	N	Modified component of the combined standard uncertainty for the fluctuation of the measured differential pressure Δp using the wind tunnel method extended
$u^*_{F, F_j, \mathrm{WTMext}}$	N	Modified component of the combined standard uncertainty for the accuracy of the force measurement at the flat belt dynamometer using the wind tunnel method extended
$u^*_{F_{\mathrm{corr}}, F_j, \mathrm{CDM}}$	N	Modified component of the combined standard uncertainty for the corrected road load using the coastdown method (after polynomial regression)
$u^*_{F_{\mathrm{fluc}}, F_j, \mathrm{AM}}$	N	Modified component of the combined standard uncertainty for the fluctuation of the force measurement in x direction using the AEROLAB extended
$u^*_{F_{\mathrm{fluc}}, F_j, \mathrm{WTMext}}$	N	Modified component of the combined standard uncertainty for the fluctuation of the force measurement at the flat belt dynamometer using the wind tunnel method extended
$u^*_{F_j, \mathrm{Brake}, F_j, \mathrm{AM}}$	N	Modified component of the combined standard uncertainty for the determination of the residual brake forces at the flat belt dynamometer using the AEROLAB method

Continued on next page

Symbols	Unit	Meaning
$u^*_{F_{j,\mathrm{Dyno}},F_{j,\mathrm{WTMext}}}$	N	Modified component of the combined standard uncertainty for the reproducibility of the road load determination at the flat belt dynamometer using the wind tunnel method extended
$u^*_{F_{j,\mathrm{WT,woB}},F_{j,\mathrm{AM}}}$	N	Modified component of the combined standard uncertainty for the repeatability of the corrected road load measured in the AEROLAB wind tunnel using the AEROLAB method
$u^*_{F_{\mathrm{TOM}},F_{j,\mathrm{WTMext}}}$	N	Modified component of the combined standard uncertainty for the accuracy of the drivetrain losses determination using TOM at the flat belt dynamometer using the wind tunnel method extended
$u^*_{F_{x,\mathrm{acc}},F_{j,\mathrm{WTMext}}}$	N	Modified component of the combined standard uncertainty for the accuracy of the force measurement in x direction using the wind tunnel method extended
$u^*_{F_{x,\mathrm{fluc}},F_{j,\mathrm{WTMext}}}$	N	Modified component of the combined standard uncertainty for the fluctuation of the force measurement in x direction using the wind tunnel method extended
$u^*_{F_{x,\mathrm{rep}},F_{j,\mathrm{WTMext}}}$	N	Modified component of the combined standard uncertainty for the repeatability of the force measurement in x direction using the wind tunnel method extended
$u^*_{F_{x,\mathrm{res}},F_{j,\mathrm{WTMext}}}$	N	Modified component of the combined standard uncertainty for the resolution of the force measurement in x direction using the wind tunnel method extended
$u^*_{f_0,F_{j,\mathrm{AM}}}$	N	Modified component of the combined standard uncertainty for the uncorrected road load coefficient f_0 using the AEROLAB method
$u^*_{f_0,F_{j,\mathrm{CDM}}}$	N	Modified component of the combined standard uncertainty for the uncorrected road load coefficient f_0 using the coast-down method (after polynomial regression)
$u^*_{f_1,F_{j,\mathrm{AM}}}$	N	Modified component of the combined standard uncertainty for the uncorrected road load coefficient f_1 using the AEROLAB method
$u^*_{f_1,F_{j,\mathrm{CDM}}}$	N	Modified component of the combined standard uncertainty for the uncorrected road load coefficient f_1 using the coast-down method (after polynomial regression)
$u^*_{f_2,F_{j,\mathrm{AM}}}$	N	Modified component of the combined standard uncertainty for the uncorrected road load coefficient f_2 using the AEROLAB method
$u^*_{f_2,F_{j,\mathrm{CDM}}}$	N	Modified component of the combined standard uncertainty for the uncorrected road load coefficient f_2 using the coast-down method (after polynomial regression)

Continued on next page

Symbols	Unit	Meaning
$u^*_{m,F_j,\mathrm{AM}}$	N	Modified component of the combined standard uncertainty for the determination of the vehicle weight before and after the measurements using the AEROLAB method
$u^*_{m,F_j,\mathrm{CDM}}$	N	Modified component of the combined standard uncertainty for the determination of the vehicle weight before and after the measurements using the coastdown method (after polynomial regression)
$u^*_{m,F_j,\mathrm{WTMext}}$	N	Modified component of the combined standard uncertainty for the determination of the vehicle weight before and after the measurements using the wind tunnel method extended
$u^*_{m_1,F_{j,CDM}}$	N	Modified component of the combined standard uncertainty for the determination of the vehicle weight m_1 before the measurement using the coastdown method (after polynomial regression)
$u^*_{m_2,F_{j,CDM}}$	N	Modified component of the combined standard uncertainty for the determination of the vehicle weight m_2 after the measurement using the coastdown method (after polynomial regression)
$u^*_{p_s,F_{j,\mathrm{WTMext}}}$	N	Modified component of the combined standard uncertainty for the accuracy of the measured static pressure p_s using the wind tunnel method extended
$u^*_{p_s,\mathrm{fluc},F_{j,\mathrm{WTMext}}}$	N	Modified component of the combined standard uncertainty for the fluctuation of the measured static pressure p_s using the wind tunnel method extended
$u^*_{p_\mathrm{Amb},F_{j,\mathrm{AM}}}$	N	Modified component of the combined standard uncertainty for the ambient pressure p_Amb using the AEROLAB method
$u^*_{p_\mathrm{Amb},F_{j,\mathrm{CDM}}}$	N	Modified component of the combined standard uncertainty for the ambient pressure p_Amb using the coastdown method (after polynomial regression)
$u^*_{p_\mathrm{Amb},\mathrm{fluc},F_{j,\mathrm{AM}}}$	N	Modified component of the combined standard uncertainty for the fluctuation of the ambient pressure p_Amb using the AEROLAB method
$u^*_{p_\mathrm{Tyre},F_{j,\mathrm{WTMext}}}$	N	Modified component of the combined standard uncertainty for the adjustment accuracy of the tyre inflation pressure using the wind tunnel method extended
$u^*_{T_\mathrm{Amb},F_{j,\mathrm{AM}}}$	N	Modified component of the combined standard uncertainty for the ambient temperature T_Amb using the AEROLAB method
$u^*_{T_\mathrm{Amb},F_{j,\mathrm{CDM}}}$	N	Modified component of the combined standard uncertainty for the ambient temperature T_Amb using the coastdown method (after polynomial regression)

Continued on next page

Symbols	Unit	Meaning
$u^*_{T_{\mathrm{Amb,fluc}},F_{j,\mathrm{AM}}}$	N	Modified component of the combined standard uncertainty for the fluctuation of the ambient temperature T_{Amb} using the AEROLAB method
$u^*_{T_{\mathrm{Dyno}},F_{j,\mathrm{WTMext}}}$	N	Modified component of the combined standard uncertainty for the ambient temperature T_{Amb} using the wind tunnel method extended
$u^*_{v_{\mathrm{Wind}},F_{j,\mathrm{CDM}}}$	N	Modified component of the combined standard uncertainty for the wind velocity v_{Wind} using the coastdown method (after polynomial regression)
u_{x_m}		Standard uncertainty of the input estimate x_m
$u_{x_m,y}$		Component of the combined standard uncertainty
$u^*_{x_m,y}$		Modified component of the combined standard uncertainty multiplied with a constant value τ
$u_{x_m,F_{j,\mathrm{AM}}}$	N	Component of the combined standard uncertainty for the influencing quantity δx_m using the AEROLAB method
$u^*_{x_m,F_{j,\mathrm{AM}}}$	N	Modified component of the combined standard uncertainty for the influencing quantity δx_m using the AEROLAB method
$u_{x_m,F_{j,\mathrm{CDM}}}$	N	Component of the combined standard uncertainty for the influencing quantity δx_m using the coastdown method (after polynomial regression)
$u^*_{x_m,F_{j,\mathrm{CDM}}}$	N	Modified component of the combined standard uncertainty for the influencing quantity δx_m using the coastdown method (after polynomial regression)
$u_{x_m,F_{j,\mathrm{WTMext}}}$	N	Component of the combined standard uncertainty for the influencing quantity δx_m using the wind tunnel method extended
$u^*_{x_m,F_{j,\mathrm{WTMext}}}$	N	Modified component of the combined standard uncertainty for the influencing quantity δx_m using the wind tunnel method extended
$u\left(x_m, x_l\right)$		Covariance of te input estimates x_m and x_l
u_y		Combined uncertainty
v	$\mathrm{km/h}, \mathrm{m/s}$	Vehicle velocity/integration variable
$v_{\mathrm{high/low}}$	$\mathrm{m/s}$	Wind and belt velocity at 180 $\mathrm{km/h}$ (high) and 40 $\mathrm{km/h}$ (low) according to the test procedure for the determination of the wheel ventilation resistance
v_i	$\mathrm{m/s}$	Velocity at time step t_i
v_j, v_{j+1}	$\mathrm{m/s}$	Reference velocity point
$v_{j+5\,\mathrm{km/h}}$	$\mathrm{m/s}$	Harmonic average of the alternate coastdown velocity points $v_j + 5\,\mathrm{km/h}$ in direction A and B

Continued on next page

Symbols	Unit	Meaning
$v_{j-5\,\text{km/h}}$	m/s	Harmonic average of the alternate coastdown velocity points $v_j - 5\,\text{km/h}$ in direction A and B
$\bar{v}_{\text{Wind}}$	m/s	Arithmetic average wind velocity alongside the test track during the measurement phase, which cannot be cancelled out by alternate coastdown runs in direction A and B
v_{w}	m/s	Wind velocity
v_x	m/s	Wheel motion velocity in x direction
v_0	m/s	Velocity at the start time t_0 of the deceleration
v_∞	m/s	Inflow velocity in the wind tunnel
W	J	Work applied by the vehicle to drive
$W_{\text{D,T,Flex}}$	J	Irreversible damping work
w_{FT}	mm	Tyre width of the front tyre
w_{RT}	mm	Tyre width of the rear tyre
w_{Veh}	m	Vehicle width as defined in Standard ISO 612:1978
w_1	N	Wind correction term
X_m		Input quantity m
X_N		Input quantity N
x		Coordinate in x direction
x_l		Expected value of a quantity l
x_m		Expected value of a quantity m
Y		Measurand
y		Coordinate in y direction
z		Coordinate in z direction

Greek symbols

Symbols	Unit	Meaning
α	degrees	Sideslip angle
α_S	degrees	Slope angle of the road
β	degrees	Rotation angle of the vehicle
Δc_w		Difference of the aerodynamic drag coefficient between c_w measured at $140\,\text{km/h}$ with fixed ride height and c_w measured with ride height depending on the wind velocity
Δd	mm	Delta value due to different tyre widths and tracks at the front and rear axles
$\Delta d_{\text{FA,RA}}$	mm	Difference of the front and rear axle
ΔF_{Roll}	N	Difference in rolling resistance measured at the flat belt dynamometer and measured in the AEROLAB wind tunnel
Δp	Pa	Differential pressure
Δt_j	s	Harmonic average of the alternate coastdown time measurements in direction A and B at the reference velocity point v_j

Continued on next page

Symbols	Unit	Meaning
Δt_{ji}	s	Harmonic average coastdown time of the ith pair of the measurements at the reference velocity point v_j
$\Delta t_{\mathrm{p}j}$	s	Harmonic average of the coastdown time at the reference velocity v_j
Δv	m/s	Incremental step between two reference velocity points ($= 10\,\mathrm{km/h}$)
$^{1}/_{2}\Delta t_{n,A/B}$	s	Imprecision of the time determination due to the velocity measurement using the GPS device
$\delta\Delta p$	Pa	Influencing quantity due to the accuracy of the differential pressure measurement
$\delta\Delta p_{\mathrm{fluc}}$	Pa	Influencing quantity due to the fluctuation of the static pressure measurement
δF	N	Influencing quantity due to the accuracy of the load cells mounted at the WDUs of the flat belt dynamometer
$\delta F_{j,\mathrm{Brake}}$	N	Influencing quantity due to the determination procedure of the residual brake forces
δF_{fluc}	N	Influencing quantity due to force fluctuations during the force measurement
$\delta F_{j,\mathrm{Dyno}}$	N	Influencing quantity of different road load measurements with braking phase at the flat belt dynamometer
$\delta F_{j,\mathrm{WT,woB}}$	N	Influencing quantity of the repeatability of the corrected road load determined measured in the AEROLAB wind tunnel at different days
δF_{TOM}	N	Influencing quantity due to the torque measurement system TOM
$\delta F_{x,\mathrm{acc}}$	N	Influencing quantity due to the accuracy of the force measurement of the wind tunnel balance
$\delta F_{x,\mathrm{fluc}}$	N	Influencing quantity due to force measurement fluctuations in the wind tunnel
$\delta F_{x,\mathrm{rep}}$	N	Uncertainty due to the repeatability of the force measurement over six measurements
$\delta F_{x,\mathrm{res}}$	N	Influencing quantity due to the resolution of the force measurement of the wind tunnel balance
δf_0	N	Influencing quantity of the uncorrected road load coefficient f_0
δf_1	N/(m/s)	Influencing quantity of the uncorrected road load coefficient f_1
δf_2	N/(m/s)^2	Influencing quantity of the uncorrected road load coefficient f_2
δm	kg	Influencing quantity of the vehicle weighing system to determine the vehicle weight before and after the used test procedure
δp_{Amb}	kPa	Influencing quantity of the measured ambient pressure p_{Amb}

Continued on next page

Symbols	Unit	Meaning
$\delta p_{\text{Amb,fluc}}$	kPa	Influencing quantity due to fluctuations of the ambient pressure p_{Amb} during the measurement
δp_{s}	Pa	Influencing quantity due to the accuracy of the static pressure measurement
$\delta p_{\text{s,fluc}}$	Pa	Influencing quantity due to the fluctuation of the static pressure measurement
δp_{Tyre}	N	Influencing quantity in the rolling resistance due to the adjustment accuracy of the tyre inflation pressure
δT_{Amb}	K	Influencing quantity of the measured ambient temperature T_{Amb}
$\delta T_{\text{Amb,fluc}}$	K	Influencing quantity due to fluctuations of the ambient temperature T_{Amb} during the measurement
δT_{Dyno}	K	Influencing quantity due to the accuracy of the temperature measurement at the flat belt dynamometer
$\delta T_{\text{Dyno,fluc}}$	K	Influencing quantity due to fluctuation of the ambient temperature during the measurement at the flat belt dynamometer
δv_{Wind}	m/s	Influencing quantity of the wind velocity
δx_m		Influencing quantity x_m
δ_0	degrees	Toe angle
ϵ		Difference in energy demand
ϵ_0	m	Distance between the centre of the wheel and the resulting force of the pressure surface of the tyre contact area
γ	degrees	Camber angle
κ		Ratio of specific heats for air with a value of 1.4
ρ_{Air}	kg/m³	Air density
ρ_{TS}	kg/m³	Air density in the test section
ρ_0	kg/m³	Dry air density, which is defined as 1.189 kg/m³
σ		Standard deviation
σ_j	s	Standard deviation of the coastdown time
σ_{q_m}		Standard deviation of n independent repeated observations $q_{m,k}$
σ^*		Standard deviation estimated using the Student t-distribution
τ		Constant factor with a value unkown to the reader
$\Theta\left(v_i\right)$		Variable depending on v_i
Θ_0		Constant term
Θ_1		Constant term
θ	degrees	Sideslip angle of the inflow
ω	1/s	Angular velocity
$\dot{\omega}$	1/s²	Rotational acceleration
ξ_m		Possible value of quantity X_m

9 783384 255969